ECONOMICS WITHOUT TIME

Also by G. D. Snooks

DEPRESSION AND RECOVERY: Western Australia, 1929–1939
DOMESDAY ECONOMY: A New Approach to Anglo-Norman History
(*with J. McDonald*)
EXPLORING SOUTHEAST ASIA'S ECONOMIC PAST
(*edited with A. J. S. Reid and J. J. Pincus*)

Economics without Time

A Science blind to the Forces of Historical Change

Graeme Donald Snooks
Coghlan Professor of Economic History
Institute of Advanced Studies
Australian National University

M
MACMILLAN

Softcover reprint of the hardcover 1st edition 1993 978-0-333-55853-9

First published 1993 by
THE MACMILLAN PRESS LTD
Houndmills, Basingstoke, Hampshire RG21 2XS
and London
Companies and representatives
throughout the world

ISBN 978-1-349-39050-2 ISBN 978-0-230-37381-5 (eBook)
DOI 10.1057/9780230373815

A catalogue record for this book is available from the British Library.

Copy-edited and typeset by Povey–Edmondson
Okehampton and Rochdale, England

To Loma Graham for time shared

Contents

PART II ECONOMICS IN TIME

List of Tables

List of Figures

Preface

This is a book about real time in economics, a dimension lost to the leading edge of the profession. This loss, it is argued, has serious implications for the role assumed by economics as the premier policy-advising body for national governments and international organisations. It is also a book about the great waves of economic change that are surging out of the distant past and into the future – waves of change that economists have failed to identify, let alone analyse. This failure has created an intellectual vacuum that natural scientists are currently attempting to fill. It is a book, therefore, that challenges economics to put its house in order before it is engulfed by this rising tide. But the question is, will economics have time?

This book is a stepping stone from the past into the future. It is part of a research programme that I began with *Domesday Economy* and which will proceed beyond *Economics without Time* to a study of forces that can be expected to influence the future of human society. Underpinning this series of books is work I am conducting on a detailed quantitative study of Britain and Europe over the last millennium. The wider research programme had its origins during study leave in London in 1979 when I began to read and think about the directions in which human society appeared to be moving. It seemed to me at the time that human society was dominated by longrun processes of economic change that stretched back into the distant past and that would continue into an equally distant future. Economics appeared to have little to say about these processes or about the forces that determine them. I came to the conclusion that to see into the future one must look into the past. To do so it seemed essential to embark upon a quantitative study of the longrun. As the best statistical source anywhere in the world for this purpose was Domesday Book compiled by order of William the Conqueror of England in 1086, and as it was part of my leisure-time reading interest in Anglo-Saxon history and language, it became my point of departure. Owing to the remarkable wealth of the Domesday data, I spent longer in the eleventh century than anticipated. But, with the basic work on Domesday Book completed, it was time to move forward in time.

Along the way I have accumulated many debts. The collaboration of John McDonald in the mid-1980s strengthened the econometric foundations of *Domesday Economy*. Since then many have listened patiently to my ideas and arguments and have called my attention to unresolved problems. I am grateful for their encouragement and advice. The chief amongst these are Paul David, Mark Elvin, Tim Hatton, Bob Jackson, Paul Johnson, Peter Lindert, Angus Maddison, Avner Offer, Barry Supple and Anthony Waterman. Sir Richard Stone kindly allowed me to read some of the draft chapters of his forthcoming book on the foundations of the empirical social sciences, and graciously discussed some of my earlier ideas. Many others have provided helpful comments in seminars and conferences in Australia, Belgium, New Zealand, and the UK. None, of course, is responsible for any heresies the book contains.

Authors always owe a debt of gratitude to those who assist with research and with the preparation of manuscripts for publication. At ANU I am particularly fortunate in having a skilled and dedicated departmental team. In particular I am grateful to Barry Howarth, Lynelle Moon, and Wayne Naughton for research assistance; Jane Berkley and Barry Howarth for copy-editing; and Ann Howarth and Barbara Trewin for wordprocessing and, even more importantly, for organising my time so that this book could be written.

Finally I wish to thank the editors of the *Australian Economic History Review* and the *Economic History Review* for permission to use material from my articles 'Rational calculation' and 'Cliometric analysis' (with J. McDonald), together with Phillimore and Co. Ltd, for permission to reproduce an extract (Figure 5.1) from Wiltshire Domesday; and Cambridge University Press for permission to reproduce a map (Figure 5.3) from H. C. Darby, *Domesday England*. All quotations from Milton can be found in *Paradise Lost* (1667) and *Paradise Regained* (1671) written, appropriately, at the end of the six centuries of statistical wilderness that stretch between the national accounts of William the Conqueror and those of Gregory King.

Sevenoaks G. D. SNOOKS
Canberra
September 1991

Introduction

Wherein past, present, future he beholds

(John Milton, 1667)

THEMES

The central themes in this book involve two interrelated ideas. The first idea is that the divorce between deductive theory and empirical reality in economics, together with the inability of theory to address, let alone resolve, the big issues facing human society in the present and the future, pose critical problems for a discipline that claims to possess social relevance. The second idea, which has considerable quantitative support, is that over the past millennium, human society in Europe has been driven through time by great waves of economic change of some 150 to 300 years in duration. Just as the forces underlying these great waves have been important in the past, they will be important in the future. Without the dimension of time in economics these very longrun forces would remain unknown. While these issues will be dealt with in detail in later chapters, they are explored here in a more general way because of their relevance to the social sciences as a whole.

Ideas and reality

Casual observation suggests that, throughout the social sciences, abstract ideas are either more 'respectable', 'successful', or 'popular' than empirical accounts of reality. Why should this be so? The answer throws modest light upon the problems facing the use of the social sciences to understand reality and it may prevent society from taking the least optimal development path. The primary objective of the social sciences is to understand man in society, in the past, present, and future. Its conduct is based upon the premise that there is an objective reality that can be explored, albeit approximately; understood, however imperfectly; and modified with difficulty. The major problem we face is that the social activities of mankind are extremely complex, interactive, and overwhelmingly difficult to understand through observation alone. In order to make the intellectual task easier to handle within the social sciences, the main activities of man in society – economic, social, and political – have been isolated and subjected to separate study; simple models of the various aspects of human society have been developed; and techniques for handling empirical evidence have been standardised. These abstract models serve a number of functions: to sidestep the complexity of social reality by developing persuasive 'stories' about cause and effect; to explore social reality so as to create order out of apparent chaos; and, in conjunction with historical analysis, to formulate 'social' policy. Theory, then, constitutes the necessary (but not sufficient) tools to assist in the understanding, and modification, of the present social reality.

The history of the social sciences has involved an intense struggle between those who think that theory should be developed inductively from observations of the past and present using historical and statistical techniques, and those who believe that it should be developed deductively with the aid of mathematics. In this struggle, the inductionists had the upper hand until the mid- to late nineteenth century, but thereafter the deductionists, through their impressive development of abstract ideas – particularly in the discipline of economics – began to surge ahead with, in Miltonian language, 'blind zeal' until they ultimately eclipsed the inductionists. The success of the deductionists was so overwhelming that the social sciences have become distinctly unbalanced in their approach to reality. Indeed, there are even times when deductive theory is mistaken for – and certainly preferred to – reality. In other words, in some circles, theory in the social sciences has come to be regarded as a better guide to reality than direct empirical studies of man in society. Policy in the social sciences, for example, is normally derived directly from simple theoretical models with little empirical analysis and no historical insight. Apart from being a precarious procedure, it is highly irresponsible, as human welfare is at stake. Social scientists, however, are not alone in this; ecologists, who are determined to remodel human society even more dramatically (despite their lack of expertise in the social sciences), also base their policies on unverified deductive models.

But what is it that accounts for the dominance of ideas, many of them unsubstantiated, in the social sciences? My general argument, which has been applied in more detail to the economics discipline in Chapter 1, begins with the fact that the social sciences attract scholars of two extreme types: those who prefer to explore worlds of their own making – the 'gameplayer' in theoretical disciplines or the 'mythmaker' in history; and those who prefer to explore the real world – the 'realist'. Naturally there are rare individuals who can work brilliantly at both ends of the spectrum, such as W.S. Jevons, the English neoclassical economist, and there are a few more who feel happy in the role of competent all-rounder; but the great majority tend to specialise at either end of the spectrum. Needless to say, a sound social science requires the application of both approaches: theory is required to make sure that the right questions are being asked, and to isolate potentially relevant relationships; while empirical/historical work is essential to test the relevance of theory and to suggest modifications to it, to suggest general causal relationships, and to provide a sound basis for policy advice. But even more importantly, the role of history is to raise and analyse the big issues that lie beyond the scope of economics – issues concerning the forces that determine the very longrun processes of economic change that are sweeping human society out of the past and into the future. Economists merely focus upon the ripples that briefly flit across the surface of these great waves (of 300 years or more in duration) of economic change. But the tools of economists are essential to historians who are attempting to analyse these forces. The exclusion of either approach, therefore, will lead to the impoverishment of the social sciences. Hence the fact that the triumph of deductive theory has marginalised empirical/historical work in the social sciences is cause for considerable concern. This concern, as it relates to the discipline of economics, is a central message of *Economics without Time*.

The imbalance between ideas and empirical work in the social sciences, which has occurred rapidly during the last hundred years, owes much to the fact that theory in the social sciences is viewed both as more intellectually respectable and more effective in maximising publications, and hence the reputation of the social scientist, than is empirical work. The academic journals in economics, for example, are dominated by minor theoretical issues that appear to have little or no relationship to reality. In many cases this work is no more than an intellectual exercise; no doubt quite 'clever' but, none the less, a mere game. This is elaborated in greater detail in Chapter 3. Of course, this approach to economics has created a long history of criticism but, as is shown in Chapter 2, it has not caused economists to change the nature of their game, because economists, like the subjects of their study, are economic rationalists. This should not surprise us. But also like the subjects of their study, economists do not have perfect foresight. They do not appear to have seen the writing on the wall of time.

The case of history is a variation upon this theme. Historians do not construct theory, but they do develop general interpretations of, and tenacious attitudes towards, periods and people that sometimes have little to do with historical fact. This is mythmaking and not history. While much of this type of 'history' is written without the realisation that it is misleading, possibly because a major data source has been overlooked, some of it is 'history' with a purpose, possibly even a noble purpose, such as social reform. The former needs to be corrected, as and when 'new' data sources or new methods become available, and the latter must be condemned as the perversion of history. Two important examples of the former 'mythmaking' that will be taken up in Part II of the book concern the view that medieval man is 'moral' man rather than economic man; and that the emergence of industrial capitalism made life worse for the mass of the population and is thus to be regretted. While these are two old historical chestnuts, neither accords with reality nor amounts to anything more than mythmaking. In Chapter 6 it is shown that medieval man did, in fact, act as if he were economically rational; and Chapter 7 demonstrates that the modern phase of economic growth, which began with the Industrial Revolution, unlike the surprisingly rapid growth that took place over the previous 600 years, massively shifted the distribution of income in favour of the lower socioeconomic classes. But these are powerful myths and they continue to attract attention even in the face of overwhelming evidence to the contrary. Why? Basically because mythmaking is easier and more sensational than reality-reconstruction; it provides the historian with a larger audience; and it is more effective in persuading others to change the system according to the mythmaker's wishes. Reality, on the other hand, is difficult to understand; it is less attractive, less popular, and is often not at all supportive of the way we wish to change the world.

Mythmaking is not confined to history. It afflicts most groups that wish to change the world, whether they are social scientists – political scientists, sociologists, economists – or natural scientists – environmentalists and ecologists. Indeed, the current attempt by ecologists to convince society that it needs to undertake a massive social, economic, and political transformation employs mythmaking – which often resembles science fiction – very effectively through the mass media and in popular publications.

The forces of change

There is a general perception amongst historians, economists, and ecologists that human society may have changed in the pre-modern period, but that it did not experience a sustained increase of GDP per capita over long periods of time. Although there are some notable exceptions, the conventional wisdom is that sustained economic growth is a modern invention. Many reasons have been advanced for this, but essentially they depend upon the way scholars view the nature of man. The general view of historians is that decision-makers in the past have been dominated by rules of custom rather than being responsive to individual self interest. Even the economists who have given any consideration to the distant past – mainly the classical economists – have convinced themselves that those who could have made rational economic decisions (the incipient middle classes) were dominated by landlords (rent seekers) who were only interested in living riotously upon the surpluses they extracted from the oppressed majority.

This view of pre-modern human nature has been able to flourish only because of an apparent lack of hard evidence needed to test the hypothesis. Instead of undertaking the necessary empirical work, economists have argued with impeccable logic, although faulty assumptions, that modern growth rates could not have been achieved over sustained periods in the past, and hence growth was not a natural condition of pre-modern human society. It has been said, for example, that: 'the simplest compound interest calculation shows how minimal the rates over long periods must have been'.[1] Of course, it is true that extrapolating modern rates of growth of GDP per capita back in time soon erodes the absolute level of average GDP until it is not capable of supporting human life – but not as soon as economists have claimed because of a number of careless assumptions they have made. These assumptions are that average income is a good proxy for average consumption, and that the functional distribution of income in the distant past is similar to that of today. In fact, during the millennia before 1700, growth rates for average consumption were very slow, whereas those for average non-consumption income were relatively high. Hence sustained rates of growth of GDP per capita for the pre-modern era, when played backwards, do not, in fact, erode consumption levels as quickly as economists have assumed. Indeed, as will be shown in Chapter 7, the compound rate of growth of England's GDP per capita for the 600 years before 1700 was as high as 0.3 per cent per annum, which implies a factor increase of 5.8 for total per capita income; but when the changing functional distribution of income over this period is taken into account, it implies a factor increase as high as 18.5 for average non-consumption income and as low as 3.0 for average consumption income (see Table 7.6). In other words, much of the growing surplus was being invested in fixed assets, as is clear from documentary evidence. Once the differential growth rates for the various components of GDP per capita are recognised, the possibility of a very longrun growth rate of 0.3 per cent per annum can be accepted. And as we know that there were long episodes of economic downturn and stagnation during this period, owing to the effects of invasion and pestilence, it is not difficult to accept the possibility that growth rates during shorter periods, of say 100 or so years, must have exceeded 1.0 or even 1.5 per cent per annum.

The recognition that the economists' diverting game of running 'history' backwards is flawed opens up an entirely new vista of human history down through the millennia to the present and into the future. The insight is simple but breathtaking in its implications: rapid economic growth for episodes of up to 300 years in duration appears to be a normal condition of human society. And it is possible that without major economic 'collapse' – which must be distinguished from Malthusian crises – very longrun growth rates would have been significantly higher than those recorded in Chapter 7.[2] This opens the way for a view of the growth of human society proceeding systematically (but not regularly) in great waves of 300 or more years in duration. Major economic 'collapses' are those caused by external crises involving invasion, pestilence, and natural disaster, and/or internal crises generated either by destructive competition or by constricting conformity – the extremes of chaos and order. But it is interesting that in Europe over the last millennium, economic 'collapse' – as occurred in 1347–49 when the size of the European economy was halved by the Black Death – did not negate earlier economic achievements but merely delayed the resumption of economic change. Economic life was not, as most scholars have thought, a matter of either endless stagnation or random zero-sum fluctuations. This general view of the progress of human society is derived from a quantitative study, in Part II, of economic growth in England during the last millennium.

CONCEPTS

This book uses terms that will raise the hackles of philosophers – terms such as reality, real world, analytical time, real time and so on. In using terms of this nature I have taken the commonsense view of the practitioner, rather than the philosophical view of the historian of ideas. These terms have functional value both as commonly understood concepts and as rhetorical devices. And in a more pedantic way I have insisted upon using one word for 'longrun', on the grounds that it refers to a well-established concept, and on reserving 'growth' for increases in real GDP per capita and 'expansion' for increases in real GDP that do not involve increases in average income.

I hold the commonsense view that there exists an objective social reality involving complex interactions between agents of change who operate within an institutional setting – involving economic, legal, political, and social dimensions – and who respond to systematic and identifiable incentives generated within this system. Over time, these social agents, by interacting with their physical and institutional environment, change the shape and nature of their society. This social reality can be explored by the human intellect, both by establishing formal systems of thought and by the critical examination of evidence generated by that reality. Clearly, our knowledge of social reality involves a continuous interaction between abstract ideas and historical evidence. The problem, as I see it, is when this interaction becomes seriously constrained – when those scholars generating abstract ideas about society limit their own access to real-world evidence, and when empiricists turn aside from theory. Economic theory is guilty of the former and general history of the latter. Of course, this is all a matter of emphasis and balance, because abstract ideas cannot exist in an intellectual vacuum, and empirical work cannot proceed without ideas. My point is that a formal process of interaction between ideas and empirical work will provide a better, although not perfect, view of social reality.

Time is introduced into this work in a number of ways. An attempt has been made to distinguish between real time and analytical time as being the primary concerns of the historian and the economist respectively. Real time refers to the sequence of events that occur in human society as viewed by historians. Real time is associated here with historical processes, and as such it is a concept in the minds of historians who are attempting to understand changes in social reality. Analytical time is an abstract idea in the minds of economists, which is related not to the unfolding of a sequence of events, but to the fulfilment of certain theoretical requirements. For example, the short-run is defined by economists as a 'period' during which fixed factors of production cannot change. Time is also used here as a rhetorical device. Historians are poets as well as scientists.

OBJECTIVES

The main objective of this book is not only to demonstrate that, in its current state, deductive economics has a very fragile grasp upon reality, but also to suggest that it is in very real danger of being toppled from its premier policy-advising role in national and international spheres. A major challenge is being mounted by the environmental scientists who are convinced that economics and economists constitute a major threat to the ecology of this planet and hence to the future of life upon it. The challenge is potentially very serious because it is riding upon a surging wave of popular support that, despite temporary setbacks, may eventually become overwhelming. And this has adverse implications not only for the economics profession, but also for society, because the environmental scientists have little or no expertise concerning man in society and because their outlook and their policies are oriented not towards human society, but to the wider ecology.

The book, then, is a call to economists to put their house in order, before they are overtaken by the tidal wave of environmentalism. This can be achieved by economists restructuring their priorities and their discipline in a more balanced way in order to gain a better perspective on reality. It is a call not for the rejection of deductive economics, but for a *rapprochement* between theory and history in the education and research of economists. It is a call to bring back real time into economics and into the formation of economic policy. I am under no illusion about the reaction of theoretical economists to this call, but they ignore it at their own peril. Earlier challenges by historicists in the late nineteenth and early twentieth centuries were briefly debated before being forgotten. It was possible to do so then because there was no external threat to economics, at least not until the last decade of the twentieth century. Ironically, time was on their side. It is no longer.

The book is divided into two parts. The first, 'Time in Economics', attempts to show how real time has been used, abused, and ignored in economics in the present and the past. Chapter 1, 'Time Lost', traces the transformation of economics from an issue-oriented empirical, indeed historical, subject, to an ideas-oriented deductive science. In the process, real time was lost. Chapter 2, 'Timely Critics', discusses the historicist and environmentalist criticism of deductive economics. Chapter 3, 'The Timeless Approach', elaborates how economics, without the benefit of historical analysis, attempts to understand the present and to proffer policy advice. The focus in this chapter is on growth theory and its limited application to reality. And Chapter 4, 'Custodians of Real Time', is a critique of economic history as it has developed in Britain, North America, and Australasia over the last century.

Part II of the book, entitled 'Economics in Time', attempts to show in a tentative way how theory, statistics, and history can be combined to provide a more realistic and relevant analysis of the process of change in Britain over the last millennium. It is shown that the comparative advantage of economics lies in analysing static outcomes, while that of history is in analysing the process of change. An attempt has been made in these chapters to expose the forces responsible for the process of very longrun growth underlying the continuous transformation of human society. It is these big issues, which are vital to the future of human society, that economics cannot cope with. But history can. Chapter 5, 'Economies "Lost" in Time', focuses upon the possibility of quantitatively reconstructing the nature of economies as early as the eleventh century – of recovering for the economist apparently 'lost' economies. Chapter 6, 'Motivation Throughout Time', attempts to test the claim of most historians that medieval man was moral man rather than economic man. The results are important, not only for our understanding of the transformation of Europe from feudalism to mercantile – and later industrial – capitalism, but also to evaluate how universally relevant is deductive economic theory. Chapter 7, 'Accounting for the Very Longrun', provides a quantitative analysis of the economic growth of England over the last 1000 years. In the process it has been possible to show, despite a conventional view to the contrary, that the English economy and, by implication, the European 'economy', was dynamic and innovative rather than static and unenterprising. It also casts light on the significance of the Industrial Revolution, which has traditionally been seen as the beginning of rapid and sustained growth, and on our ability to maintain growth in the future. Finally, Chapter 8, 'Time Regained?', discusses the future role of economic history within the discipline of economics.

OUTCOMES

The chapters that follow present a variety of new empirical conclusions and ideas: an explanation of the increasingly abstract nature of social sciences in general and economics in particular over the last two centuries; the evolving instruments of economic policy over the past 2000 years; the challenges currently facing economics; the discovery that economic history is not the offspring of economics and history and should be thought of as an older sibling; the relationship between the various economic history traditions in the English-speaking world and particularly the pioneering role of the Australian analytical tradition; the innovative and dynamic nature of feudal and medieval economic systems; the fact that medieval man was economic man, not moral man; the underlying processes of change in human society which have a momentum that links past, present, and future; the suggestion that human society over the past millennium has been dominated by great waves of change of 150 to over 300 years in duration; and the inability of economics to identify, let alone analyse, these very longrun forces that are critical to the future of human society.

As these conclusions and ideas challenge the conventional wisdom in a variety of fields, they will inevitably be vigorously opposed, not only for their content and implications, but also for the unapologetic way in which they have been presented. It seems to me not only, as T. S. Eliot has said, that 'Old men ought to be explorers',[3] – together with those who are not so old – but that new ideas should always be clearly distinguished from the ideas that they are attempting to displace, rather than being surreptitiously attached to an existing body of conflicting knowledge. In this way new ideas can be openly debated, and either accepted or dismissed on their merits. Otherwise, new ideas will be deliberately ignored, unwittingly added to a stock of inconsistent knowledge, or resisted in a vague and non-specific way. An interesting example of the confusion created by the conventional wisdom attempting to balance mutually exclusive ideas is the different views concerning the nature of danegeld held by F. W. Maitland and J. H. Round in the 1890s. Round, as will be shown in Chapter 6, claimed that the tax system was arbitrary, whereas Maitland thought that it had a systematic relationship to manorial income. Instead of drawing out the differences between these conflicting ideas, Maitland attempted to disguise them. As a result, Round's empirically incorrect view prevailed for the next century, and provided support to those who regarded agents in the economic system of feudalism as economically irrational.

Something of what I mean, although in a different context, was recently stated by a philosopher of knowledge. Robert D'Amico has written:

> Feyerabend argues that in establishing or defending a new scientific theory the defender is never confronted with specific or discrete objections but with a total resistance to the new scheme. Conversely the revolutionary must treat the old theory as wrong 'root and branch'. The factual data from an older tradition acts as a 'mental block' against change because old data, observation, and reasoning is 'contaminated' with such obstacles as the 'obvious' or what is 'natural'.[4]

This is largely true as far as it goes. New ideas are resisted because of entrenched ways of thinking, but, more fundamentally, they are resisted because they are perceived to challenge established reputations and render existing human capital obsolete. The outcome of this resistance varies according to whether the intellectual tradition is open or closed, which in turn depends upon the competitive ability and confidence of its members. If the intellectual system is open, then the resistance to new ideas will create a competitive exchange which will advance knowledge, even if the challenging idea is eventually proved to be wrong; but if the intellectual system is closed, then the new idea will be buried, even if it is correct.

The question we now turn to is: does economics have a future? It is possible that, with determined goodwill on all sides, a reconciliation could be achieved between the estranged parties, and that time, which will provide insight regarding the big issues facing human society, could be regained by the economics profession. Time, however, is running out.

Part I

Time in Economics

1 Time Lost

The flower of heaven, once yours, now lost

(John Milton, 1667)

THEORY AND HISTORY

The only justification for any discipline in the social sciences, apart from its educational value, is whether it assists practical men and women, together with policy-makers, to understand the reality of human society. For economists, that reality is the process by which human beings attempt to satisfy their material needs and desires.[1] Accordingly, the tools of analysis – the deductive theory – should not be confused with the reality that is being pursued. As will be suggested in later chapters, there is a subtle tendency, despite the protestation of some economists, for such confusion to arise. It is this tendency – a result of the short cut to success by economics – which threatens to undermine the position that the discipline has achieved. In these circumstances there is good reason for economists to pause and reflect upon what they are doing and why. It is in this spirit that the following chapters draw attention to the increasingly fragile grasp that economics has on reality, and suggest how that reality can be regained. This is meant to be an exercise of constructive criticism from within the profession. There are those from without who are determined to displace economics entirely, with a prescriptive discipline based upon the natural sciences.

Briefly, the argument in this chapter is that the emergence of economics as a deductively-based discipline, particularly during the last hundred years or so, has led to the downgrading of the importance of applied economics in general and economic history in particular. Over this time, the discipline has been transformed from the practical empirical art of political economy to the abstract deductive science of economics. While this methodological transformation was necessary for the development of the discipline, it brought with it significant costs, particularly the loss of the applied and historical skills that are essential to the exploration of reality. These costs resulted from the tempting short cut taken by the discipline in its rush to the forefront of the social sciences. Had a balance been maintained by the profession between abstract deductive theory on the one hand, and applied/historical analysis on the other, its position today would be more secure. Unfortunately, economics has developed in an unbalanced fashion during the past century, and the theorists – or gameplayers as I will call them here – rather than the realists have come to dominate the profession. As a result, economics has increasingly come to reflect intellectual fashion rather than the reality that alone can justify its influential position in modern society. Of course, there have been exceptional economists, such as Richard Stone, who have attempted to bridge the widening gap between theory and reality, but the profession has not followed their lead.

It is further argued that economic history, broadly defined to encompass all human time down to the present, is the laboratory of economics.[2] Experimental economics, which has recently attracted greater interest in the profession, is not a substitute for economic history because its scope of enquiry is limited to issues of motivation and static neoclassical market models, and these only in the present. As history is the only method of exploring the full scope of reality directly,[3] the loss of the skills of historical analysis by the discipline – the loss of real time – has led to a receding reality in economics. The implications are critically important. It is argued that the loss of historical, or longrun, perspective is a major reason for the growing irrelevance of a large proportion of economic theory and for the failure of economic policy in Western economies during the 1980s – a failure that has led to, or at least exacerbated, high rates of unemployment and inflation, dangerously large deficits on current account, unprecedentedly high interest rates, excessive speculation, a heavy dependence upon overseas capital and technology and the growing failure of key financial institutions.[4] But even more importantly, economics has failed to recognise, let alone analyse, the underlying forces of economic change that are responsible for transforming human society. The problems arising from these profound and enduring forces are likely to dominate the immediate future.

By advancing these claims, I am not advocating the rejection of deductive economics. On the contrary, deductive economics has an essential role to play, not only in suggesting possible economic relationships that may exist in reality, but also as an aid to historical analysis. Rather, my argument is for a more balanced approach to economic studies, an approach that will emerge from a *rapprochement* between theory and history. Theory, history, and statistical method must become equally important foundations for the teaching and practice (particularly regarding policy) of economics. Theory must be refined by history, just as history must be informed by theory.[5] Economics began as a practical empirical art and, although it has acquired important and powerful tools of analysis, it cannot afford to lose itself in the world of abstract ideas, where it is in very real danger of becoming a branch of metaphysics and hence being irrelevant to the practical world of mankind. It is the purpose of this chapter to explore the process by which, in the profession of economics, time was lost.

THEN AND NOW

A stark contrast is provided by comparing the economics profession at the time of its emergence as a 'modern' social science in the late seventeenth and early eighteenth centuries, with its position in the late twentieth century. In the time of Petty (1623–87), Davenant (1656–1714), King (1648–1712), Law (1671–1729), Cantillon (1680–1734), and Steuart (1712–80), the newly emerging discipline of political economy attempted to emulate the natural sciences by developing generalisations or principles based upon the application of logical thought to the empirical results obtained from a systematic observation of reality. In other words, it attempted to adopt a 'Newtonian experimental, inductive method of inquiry'.[6] The empirical data employed by the early political economists included both historical observation and relatively sophisticated estimates of national income – the so-called Political Arithmetic of Petty, Davenant, and King.[7] This evidence was collected and reworked in order to resolve urgent economic problems and to propose corrective action. In the process, *ad hoc* generalisations that could explain what was happening in the real world were developed. At this early stage, however, the political economists were unable, or thought it undesirable, to follow Isaac Newton's lead in expressing these generalisations in mathematical form. The historical–inductive approach of early political economy should not come as a surprise. As will be argued in Chapter 3 ('Policy'), modern economics emerged from an intellectual tradition stretching back beyond the ancient civilisations of Rome and Greece in which current economic problems were explored, and policy formulated, using an historical–inductive methodology.

By the early eighteenth century, therefore, the new 'discipline' of political economy had produced a body of literature – in which the principal actors were motivated by self interest – that attempted to explain and predict the working of aspects of the English economy: aspects that had urgent policy implications. While this literature presented itself as objective and 'scientifically' based, it had yet to break free from a moral and philosophical framework. In the beginning, the 'discipline' of political economy was a pragmatic, shapeless mixture of *ad hoc* inductive generalisation, current empirical data, and historical information, all of which was overlaid with moral and philosophical judgements. The later specialist academic areas of theory, statistical analysis, and economic history had yet to emerge from this primeval chaos.

Looked at from the end of the twentieth century, the discipline of economics has been much changed by the intervening 300 years. Indeed, the pioneering political economists would find it difficult to recognise the discipline as the one they had brought to life. Although considerable emphasis is still given to the policy orientation of the discipline, a large proportion of economists spend most of their careers acting as if the important issues are intellectual rather than practical. The correlation between the content of leading economic journals and learned conferences on the one hand, and urgent economic problems on the other, does not appear to be very high – a fact that has not gone unnoticed by interested observers, such as financial journalists and concerned economists.[8] It is quite clear that the agenda for much of the content of academic journals is determined from within rather than from without the profession. Economic theory feeds upon itself at least as much as upon real-world problems. In many cases, either no pretence is made to provide solutions to real-world problems, or the 'solutions' provided merely concern artificial problems.

A major difference between 'then and now' is that the previously undifferentiated field of economic studies has become divided into a number of broad areas between which there is little communication. The economics profession now consists of three main subgroups: the theorist, the applied economist (who is led by the econometrician), and the economic historian. Of these subgroups, the most prestigious within the profession are the theorists, followed by the econometricians and other applied economists, and last, and certainly least, by the economic historians. In each of these divisions the elite groups tend to be those most skilled in technical abilities such as mathematics in theory, statistical methods in applied economics, and theoretical and quantitative techniques in economic history. Accordingly, in each area the emphasis has been upon the sophistication of technique rather than upon the understanding of reality. The fundamental weakness of the profession is the refusal to integrate these broad divisions into an effective attempt to understand reality. It would appear that the internal dynamics of the profession are centrifugal in nature, and it will need an exogenous shock to fashion the various parts into a new and more effective structure.

The really fundamental contrast, however, concerns the discipline's methodology. The historical–inductive method of the early political economists has yielded to the ahistorical–deductive method – a method that is highly technical and mathematical – of modern economists. No longer are the principles of economics drawn from systematic observation of past events – the exploration of reality – but rather from a process of logical deduction based upon intuitive, or even arbitrary, assumptions. Modern economic theory, therefore, is an intellectual game rather than the generalised explanation of an observed reality; it is determined more by the world of ideas than by the world of human affairs. No doubt the pioneers of political economy would be perplexed by the course taken by economic theory.

Ironically, the increasing precision and abstract nature of economics has been accompanied by the profession's growing status and influence. The less it is understood by the layman the more it is respected. Prior to the mid-seventeenth century the economic policy and intervention of governments were determined by the practical experience and the historical understanding of the ruling elite and their agents. As shown in Chapter 3, the past was the only guide to the present, and the economy was run as if it were the personal estate of the ruler.[9] With the continued emergence of markets and the growing complexity of economic activity throughout the medieval and early modern period, the English economy could no longer be treated simply as a large estate. Greater economic management skills, together with a better understanding of underlying economic processes, were required. The emergence of political economy, which is usually examined only from the supply side,[10] must be seen as a response to these changing requirements. Political economists such as William Petty, Gregory King, David Ricardo and Thomas Malthus, together with policy advisers influenced by the new science, gained a debating advantage over their lay colleagues. In part this was due to the greater clarity with which they saw economic processes, and in part to the intimidating technical armoury they possessed.[11] To compete with the economically literate, other policy advisers had to become familiar with the basic principles of political economy. Thus the profession grew and, at the same time, became increasingly remote from the uninitiated. Indeed, the more intellectually rigorous economic theory became, the more it attracted intelligent and ambitious students who wished to employ it in the pursuit of academic careers, or careers in business and public service.

The largely incorrect association of professional economic advice with economic prosperity also played an important part in the discipline's development. As economists became more involved in advising decision-makers in government and business they became closely identified with the rise and fall of economic prosperity. When the economy was buoyant, and those who held political power were prosperous, the status of economists rose and their ranks increased. Of course, the acclaim given to economists is out of all proportion to their real role in creating and sustaining periods of boom. Similarly, economists have had to suffer a disproportionate share of the blame for recessions and depressions, and during these periods their status and influence has declined – the 1980s were not unique. But the profession appears to gain more during the booms than it loses during the recessions. Institutional progress in the profession, therefore, appears to resemble the ratchet effect.

Our comparison of origins and present outcomes suggests, therefore, that the nature and influence of the economics profession has changed radically over the past 300 years. In the beginning, economics was an historical, philosophical, and moral enquiry, whereas it is now self-consciously ahistorical, aphilosophical, amoral, abstract, and mathematical. It has been transformed from a discipline intent upon exploring reality in order to solve urgent economic problems, to one in which the world of ideas is more important than the world of human affairs.

FROM REALISM TO ABSTRACTION

The history of economic thought is a well-worked area of intellectual endeavour. My interest is not in recounting the contribution made by the great economists over the last 300 years,[12] but rather to trace out and explain the shift in professional preoccupation in economics from the exploration of reality to the playing of abstract games. There are a number of dimensions to this transformation, which include:

- the methodological shift from induction to deduction;
- the declining status and influence of applied economics and economic history; and
- a changing focus from longrun to shortrun, together with a shift in emphasis from dynamic to static analyses.

Induction to deduction

In the beginning

The attempt by the early political economists to apply the prevailing scientific method – the Newtonian empirical–inductive approach – to exploring economic reality with a view to changing it for the better was only partially successful. Although they struggled to achieve a degree of objectivity in their work, they were unable, or thought it undesirable, to break free from the prevailing philosophical and moral climate of the time.[13] And in contrast to their colleagues in physics, they did not express their inductively-based generalisations in mathematical form, nor did they develop integrated models of the prevailing economic system. The task of developing sophisticated theory through the process of induction proved, for reasons to be discussed later, to be too difficult.

The first major steps in the long process of methodological transition were taken independently by the French Physiocrats and Adam Smith in the second half of the eighteenth century. Their contributions involved attempts to discover and model the basic underlying processes in an economy. The Physiocrats, the chief of whom were François Quesnay (1694–1774) and Comte de Mirabeau (1715–89), attempted to model the economy in a quantitatively precise way to show the circular flow of incomes and expenditures between the three economic groups of landlords, farmers, and 'unproductive' (that is, non-agricultural) workers. Although it was a rather simple and unrealistic model, which did not have many imitators, it was an early example of an abstract form of analysis arrived at by a mixture of observation and deduction.[14]

Adam Smith (1723–90) in the *Wealth of Nations* also attempted to explore the underlying economic processes, in his case by developing a model based upon division of labour and capital accumulation, to explain the process of growth that he saw as being a natural outcome of the English economy in the late eighteenth century. Using as his central assumption the now familiar idea that economic agents acted consistently in their own self interest, Smith developed simple models by employing historical observation together with inductive and deductive thought. To persuade others that these simple abstract theories (which are really quite imprecise) were capable of revealing the underlying mechanism of economic change, Smith marshalled historical and comparative data to 'illustrate' their relevance.[15] His purpose in isolating a model of the English economic system was to propose advice that would maximise the rate of economic growth – namely to work within the parameters of the system rather than to divert and frustrate it. Although Smith built upon the work of his predecessors, who attempted to analyse the English economic system that they saw as being held together by the self interest of economic agents – writers such as John Law, Richard Cantillon, and James Steuart – he established an approach to the subject that set economics on its modern course. And in doing so, in Miltonian language, 'Adam . . . lost paradise'.

Although Adam Smith's work provided a sound foundation for future developments in economics, it also initiated an historical sleight of hand. It has already been noted that he attempted to demonstrate the relevance of his theories by *illustrating* them with historical examples. I say sleight of hand because only those facts, events, processes and interpretations of history that support a particular theory are paraded, while the rest of history, which may be unaccommodating, is ignored. It was for this use of history that William Cunningham attacked Alfred Marshall in an article called 'The perversion of economic history' in the *Economic Journal* for 1892. And it is a misuse of history that is with us still.[16] It is an interesting phenomenon. Economists may apply the most rigorous econometric techniques to contemporary cross-sectional and time-series data when testing their theories, but when they attempt to provide them with greater universality by appealing to longer periods of time, they invariably abandon their rigour in favour of more impressionistic techniques. It is almost as if they believe that history is not worthy of serious study and can be plundered in a haphazard manner. As I hope to show in the second half of this book, history certainly is worthy of serious study and rigorous methods can and should be employed in reconstructing the past, even as far back as 1000 years before the present.

The first assault on real time: Ricardo and the neoclassicists

Although Adam Smith employed deductive reasoning, albeit not very precisely, it was David Ricardo (1772–1823) who completely divorced economic model building from the experience of the past. Ricardo stripped economics of everything he considered superfluous to the economic problem under consideration, including history, sociology, philosophy, and the institutional setting. He did this by abandoning the inductive–deductive approach employed by earlier political economists in favour of a consistently deductive approach based upon the central assumption of economic rationality. As a recent commentator has said: 'His ingenious mind, essentially that of a brilliant theoretician, never displayed any significant interest in the past'.[17] The resulting theory, which addressed central issues of the day, was ahistorical and abstract to the degree that it could have been (but was not) expressed in mathematical form. No doubt, as Paul Samuelson has suggested, had Ricardo employed mathematics his arguments would have been less labyrinthine.[18] Much of Ricardo's theory focused upon microeconomic issues concerning the theory of value and the functional distribution of income. It is important to realise that the abstract, ahistorical, and aphilosophical nature of Ricardo's work was unique – and was perceived to be unique – in the early nineteenth century; Thomas Malthus (1766–1834), for example, commented that: 'The principal cause of error, and of the differences which prevail at present among the scientific writers in political economy, appears to me to be a precipitate attempt to simplify and generalize'.[19] Yet while this approach was unique in Ricardo's time it was to become the standard approach in the discipline with the development of the new neoclassical economics in the last quarter of the same century. While Malthus criticised the methodology of Ricardo, the later reaction to his own work varied. Some, such as T. E. Cliffe Leslie and H. S. Foxwell, saw him as using the historical method of induction, while others, such as J. K. Ingram, felt that Malthus had begun deductively, and only in subsequent editions had added a large amount of 'illustrative' historical material.[20] The latter view is more balanced.

With the simultaneous but independent development of the marginal theory of value by W. S. Jevons (1835–82) in England,[21] Carl Menger (1840–1921) in Austria, and Léon Walras (1834–1910) in Lausanne in the 1870s, deductive economics took on an international character. The new microeconomics – the marginal revolution – that emerged in the closing decades of the nineteenth century was a response not only to the changing economic system in Europe which generated a demand for a more systematic and sophisticated body of economic principles, but also to the general intellectual climate that was becoming increasingly scientific.[22] What is important here for our purposes is not the nature of the theoretical contribution in microeconomics, but the nature of its methodological underpinning. It is generally recognised that this period marked the watershed between political economy and economics – between, in other words, the art and the science of the discipline. In this process of transformation, economics – or, more to the point, neoclassical economics – became more abstract, deductive and highly simplified, and, in the process, it became less realistic. Mathematical elegance replaced the clumsiness of a science clothed in words. But the new economics was less accessible to the uninitiated than was classical economics, not only because it was more abstract and mathematical, but also because it appeared little concerned with macroeconomic issues.

Alfred Marshall (1842–1924) gave the impression of bringing intellectual balance back into neoclassical economics.[23] Not only was he elegantly able to reconcile the combatants in the game of value theory by employing the effective demand–supply model (which had been anticipated by J. S. Mill), but Marshall also gave the impression to the uninitiated that he had placed neoclassical economics back into historical context and into the language of the educated layman. Yet his achievement was more apparent than real. Although Marshall apparently intended to place his analysis of the 'present' in 'an evolutionary, hence a historical, theory of economic change' this grand design did not get very far.[24] His rhetoric and his practice did not coincide. The historical chapters in the *Principles* are – unlike his theoretical work – vague, imprecise, and tautological, and constitute an impressionistic appeal to the past. One scholar has referred to it as 'little more than historical theory'.[25] In this, Marshall is typical of economists who generally take an unprofessional approach to history. Basically he was making a half-hearted gesture to the intellectual expectations of the time. Marshall used history merely to rationalise his theory, to give it a veneer of universality and, possibly, to placate his more aggressive 'historicist' opponents such as William Cunningham (1849–1919). History was used, therefore, for personal rather than professional ends.

The initiated, however, were not deceived. The younger generation of economics students were attracted by the abstract and mathematical rigour of Marshall's theory, and recognised his fundamental opposition to the operational use of history in economics. They were required to do very little history in their courses, and what they did focused narrowly on the previous few decades: two-thirds of the new Cambridge economics degree was reserved for theory, with only one-third for applied economics, political science, and 'recent' economic history.[26] In a revealing autobiographical sketch in 1917, for example, Marshall recalled his first encounter with economics after an earlier training in mathematics: 'they [his mentors] told me that I needed to study Political Economy. I followed their advice, and regarded myself as a wanderer in the land of dry facts; looking forward to a speedy return to the luxuriance of pure thought'.[27] He also said: 'I don't much believe in the possibility of direct induction', and admitted that he only used history for 'illustrative purposes'.[28] This was the real example that Marshall provided for those who came after him: the real core of economics was pure thought, and empirical–historical material was only useful in so far as it illustrated the supremacy of pure thought. Marshall's fiercest opponent, Cunningham, saw this clearly, and did not like what he saw. In his famous 1892 article, mentioned earlier, Cunningham attacked Marshall for attempting to support his deductively based theory by an *impressionistic* appeal to the past. This approach, which can be traced back to Adam Smith and beyond, 'perverted' economic history because it extracted only those facts, events, processes, or interpretations that supported a predetermined point of view, ignoring all that contradicted it.

This is not to say that Marshall and his followers were uninterested in practical problems and in influencing the affairs of men with their abstract ideas. But in their work the problems and policies arose largely from abstract theory and an intuitive grasp of the world around them, rather than from a detailed empirical examination of reality. Although the neoclassicists adopted mathematical methods to facilitate the development of deductive theory, they (with the exception of William Jevons) had little time for statistical techniques that could have been used to apply their theory to the real world.[29] Although one might expect a considerable degree of overlap between abstract and real problems while economists keep a close check on the world around them, it is highly probable, as the balance of interests shifts from empirical to theoretical matters, that problems and policies will become subordinated to intellectual priorities. This appears to have happened over the last hundred years. Accordingly, Cunningham's charge about the perversion of economic history can be extended to the perversion of policy-making, with theory often dictating the issues rather than the reverse.

The counter-attack: the historical school

There were those who reacted against the abstract, deductive form of economics created by Ricardo and extended by the neoclassicists. The most forceful amongst these were adherents of the historical school of economics, who accused the Ricardians and neoclassicists of being dangerously abstract and unrealistic. History, they argued, provided a far sounder basis for deriving economic principles. As is well known, this empirical–inductive school of thought had its origins in Germany as a reaction against the British deductive school,[30] partly because of German unwillingness to accept the policy implications – particularly concerning free trade – of the British profession, but also partly as a romantic reaction against the excesses of reason. While romanticism pervaded most aspects of German thought in the early nineteenth century, its influence in England was largely limited to literary writers – Samuel Taylor Coleridge, for example, attacked the Ricardian school of economics[31] – and indirectly to the English historicists. It is an interesting fact that the German historical school not only rejected the abstract deductive approach of Ricardo and his followers, but in the later nineteenth century also sought methodological inspiration in the biological rather than the physical sciences. Accordingly, they emphasised the evolution of economies and were concerned with the organic unity of the process of change. More recently a cliometric historian has explored the different insights that economics might have provided had it adopted the methodology of the biological sciences.[32]

The historical school in Britain, which resisted the deductive nature of orthodox economics and the central assumption of economic rationality, derived inspiration from the German reaction against deductive theory. As discussed in Chapter 2, the two leading proponents of the British historical school in the 1870s were Irish scholars – T.E. Cliffe Leslie (1827–82), a political economist, and J.K. Ingram (1823–1907), an historian.[33] They claimed that all economic thought was relative to time and place and hence was not universal. Their influence was greater than it might have been at another time, because it was generally recognised at this stage that classical economics was more relevant to a pre-industrial society than to the mature capitalist economy that Britain had become. The influence of classical economics therefore began to recede; indeed, it probably would not have lasted so long but for the amazing intellectual influence of J. S. Mill (1806–73), a disciple of Ricardo, who attempted to make the master's work more realistic by the use of historical and contemporary illustration.[34] For a decade or so, the historical school was a serious contender for the vacant title of economics, at least in terms of providing a new basis for economic policy, if not in providing an alternative body of economic theory.[35] It was more acceptable to the majority of the profession than the brash new neoclassical school, because very few economists of the time had the necessary mathematical training, whereas most had extensive exposure to history. Indeed, at both Oxford and Cambridge during this period, economics was taught as part of the history programme.[36] The American institutional school, the most lasting memorial of which was the National Bureau of Economic Research (NBER) established in 1920, was also influenced in its anti-deductive approach (which included D. Dewey, J. M. Clark, and W. Mitchell) by the German historical school.

Yet the influence of the historical school was not to last. As N. S. B. Gras commented in 1920, 'The historical method, as applied to economic studies, seems in large part to have failed at the hands of both the younger and older group of historical economists'.[37] In the first place it had promised more than it could deliver; the historical school failed to produce an alternative structure of inductive thought that could stand alongside that of the rapidly developing deductive school.[38] The intellectual task of exploring the economic processes underlying historical reality was far more difficult than the construction of simple deductive mathematical models. Secondly, it is difficult in an inductive school to obtain agreement about a body of generalised ideas. There are a number of reasons for this. On the one hand it is well known that inductive inferences cannot be drawn precisely from the premises, but only with some degrees of probability, because there is no logical necessity for the unobserved to continue the pattern discerned in the observed. Different scholars therefore can, and do, make different inferences based upon the same number and type of observations. On the other hand, as different scholars may have access to different numbers and types of observations, they may arrive at different inferences even if they could agree on those occasions that the type and number of observations were the same. Thirdly, the deductive school in the late nineteenth and early twentieth centuries enjoyed a growing reputation both in the academic field, because it was able to provide a manageable intellectual challenge for 'bright' university students, and in the political and commercial spheres, because it provided the tools required to tackle practical microeconomic problems in the largely buoyant late Victorian–Edwardian economy, where growth was no longer an urgent issue. Fourthly, Marshall's attempt – deliberate or not – to disguise the degree of abstraction of his neoclassical synthesis and to give it an historical veneer, probably helped to deflect some of the force of the historical school's attack, and to make his methods more palatable to a wider audience. Fifthly, the success of Marshall at Cambridge in 1903 in finally breaking free of history's apron strings and in establishing a separate economics tripos, meant that students could be given larger doses of history-free economics, could be nurtured away from the glowering presence of hell-fire historicists such as the Revd Cunningham, and could be warned of the 'dangerous fallacy' of the historical school.[39] Finally, J. N. Keynes' *Scope and Method of Political Economy* was considered by the neoclassical school to be the last word on the relative merits of the deductive and inductive approaches.

Despite the resounding victory of the neoclassicists, supporters of the historical school did linger on at the London School of Economics until the late 1920s, under the protective influence of Beatrice and Sidney Webb. The leading lights at the LSE in this period, all of whom were appointed because of their anti-neoclassical sympathies, included W. A. S. Hewins, Edwin Cannan, and R. H. Tawney. It was not until the appointment of John Hicks in 1926, Lionel Robbins in 1929, and F. A. Hayek in 1931 that neoclassical economics gained the ascendency.[40] This marked the end of the English historical school as an effective academic force. Its only intellectual heirs are the economic and social historians in Britain who have faithfully carried on the anti-deductive and anti-economic-rationality attitudes of the founding fathers, Cunningham and Ashley – but not the desire to generalise from the particular, which had characterised the historical school – down to the present time. None of those historians, however, has even been able to attract the attention of orthodox economists, let alone cut them to the quick as Cunningham was able to do to Marshall a hundred years ago. Instead, they have been left to snipe ineffectively at the flanks of the analytical economic historians who employ orthodox economic theory and quantitative techniques in their work.

As it is important to understand why the historical school failed in its bid to provide an alternative body of economic principles to the classical and neoclassical schools, it is worth drawing the above ideas together in a concise way. Clearly, the explanation is complex, but it seems to me to involve two interrelated elements. The first concerns the extremely difficult, although not impossible, task of drawing generalisations from reality.[41] In other words, the task that the historical school set for itself was far more difficult intellectually than the neoclassical agenda to construct simple deductive models. This is so for two reasons. First, few would disagree that the process of induction demands considerable ability to examine the apparent chaos of reality in order to discern the underlying processes that drive it. There is always a very real danger that such ambitious intellectual endeavour will collapse under the weight of its own complexity. Secondly, as already discussed, there is the problem of obtaining agreement in an inductive discipline. In contrast, abstract deductive model building is a challenging, but more modest, task for those who are skilled in the ways of logic and mathematics. Given the limitations of the human mind, progress can be made more rapidly in the social sciences by the use of deduction rather than induction. Also, owing to the general acceptance of the philosophical basis of deduction, it is easier to obtain agreement about the body of generalisation in the deductive school.

This brings us to the second and closely related point that the historical school failed because it refused to employ the simple models of its arch-rival. The historical school made the fatal error of competing, rather than co-operating, with neoclassical economics. Its decline and eventual extinction should be seen in these terms. There is a lesson here: if the profession of economic studies is to reach its full potential, theory and history must be regarded by the great majority of its practitioners as complementary rather than competing activities. Neither partner will prosper for long without the other. When economic history rejects economic theory – as in the case of British economic and social history – it loses its identity and runs the risk of being absorbed by general history. Similarly, theory without history (that is, reality) runs the risk of losing the confidence of those who pay the piper, and hence of having continuing influence in the world.

Rolling back real time

With the demise of its opposition, the neoclassical school grew rapidly in influence. In the first few decades of the twentieth century, Marshall's *Principles* provided a stimulating intellectual basis for research and teaching in economics. Although those involved in developing economics claimed to be concerned with urgent economic problems, their preoccupation – if we ignore 'deviants' such as Joseph Schumpeter, who focused upon the process of macroeconomic change

– was with microeconomic issues. In the event, their work became even more abstract, ahistorical and, as unemployment emerged as a major problem after the First World War, distant from the concerns of policymakers. The point at which the intelligent layman could profitably read the work of economists had long been passed. During these years, Marshallian economics was amplified and extended in England (by Arthur Pigou, Dennis Robertson, J. M. Keynes, and Joan Robinson), in Europe (by Léon Walras, Eugen von Böhm-Bawerk, Knut Wicksell and so on), and in the USA (J. B. Clark) in order to analyse general equilibrium, welfare, capital and interest, and money, interest, and prices. Whatever the indirect practical impact of their work – and it appears to have been very limited – clearly they were preoccupied with pure theory.

While large-scale unemployment in Britain, together with the precarious state of the international economy in the inter-war period, led neoclassical economists to take macroeconomic issues more seriously than before, the emphasis on abstraction became more intensive and extensive in the profession. Although the remarkable theoretical contribution made by J. M. Keynes in the *General Theory*, for example, was in response to the fact of the Great Depression, it was based not upon a closer examination of historical processes, but on the intuition of that great man about current economic institutions and circumstances. In a revealing statement, Keynes said:

> It is astonishing what foolish things one can temporarily believe if one thinks too long alone, particularly in economics (along with the other moral sciences), where it is often impossible to bring one's ideas to conclusive test either formal or experimental.[42]

Here he discounts the use of history as a method of testing 'one's ideas', possibly in part because he was convinced by his father's dismissal of the methodology of the historical school[43] but more fundamentally because he was aware from his earliest academic work of the problem of induction.[44] Keynes instead appeals to the intuition of fellow theorists rather than the results of either applied economics or economic history, which was left to others – those lesser mortals who were prepared to get their hands dirty testing some of Keynes' propositions (such as the consumption function, the multiplier, and the trade-off between inflation and unemployment) using historical data. Keynes himself was quite content to move directly from abstract theory to advising governments of various nations without empirically testing his theory. Indeed, there is good evidence to suggest that he developed the policy before the theory.

Ironically, the *General Theory* not only directly stimulated empirical work; it also encouraged the official collection of macroeconomic data that, in conjunction with the development of econometrics from about the same time,[45] led to the emergence, largely in the USA, of a new industry of applied research in the neoclassical as well as the Keynesian traditions. The importance of the contribution of this applied work to the development of theory is often underrated, particularly in Britain (and there, particularly at Cambridge – Keynes' *alma mater*). As G. J. Stigler points out in his review of *The New Palgrave: A Dictionary of Economics*:[46]

> Many empirical findings of economics have histories that are at least as interesting and important as most theoretical developments. The history of the estimates of the consumption function, and the effect these estimates had on professional thinking, is one example. The estimates of the Cobb–Douglas production function, and the growing sophistication of the estimation procedures, is a second example. The influence of measurements of price levels on monetary analysis is a third example.

He concludes by saying that 'The decision to exclude empirical studies is a serious blemish'. He could also have mentioned the empirical/historical work underlying the Phillips curve and trade cycle analysis. It is significant that a major multi-volume compendium of economics that attempts to encompass the achievements of the profession both past and present, can overlook the theoretical contribution of the large body of significant empirical work undertaken since the Great Depression. Economics, particularly in England, is still firmly in the hands of anti-empiricists. One of the main reasons for the lack of integration of applied work to the body of economic theory is the separate development of the basic techniques used in both. Economic theorists adopted mathematical methods as early as the 1870s (although they did not become widespread until after the 1930s) in order to develop their deductive models, but neglected applied statistical techniques. Those techniques – which were introduced in the early years of the twentieth century, were used more widely after 1931 when the Econometrics Society was established and Richard Stone's Department of Applied Economics at Cambridge was underway, but only became common after the mid-1940s – would have provided some systematic information about the real world. The two traditions have always remained separate, possibly because, from the beginning, mathematics has been associated with deduction, while statistics, and later econometrics, were associated with induction – the method of the historical school.

If the economics profession, on important occasions such as the publication of *The New Palgrave*, can overlook the contribution to theory of its applied economics branch, what role does it see for economic history? Of the 2000 entries in *The New Palgrave* only a handful are devoted to topics and scholars in economic history,[47] and one of these topics is a six-page entry on economic history itself. Conspicuously absent is any discussion of historical method or the historical treatment of contemporary issues. How does this compare with the position when, between 1894 and 1899, Palgrave published the first edition of his *Dictionary of Political Economy*? The first edition is strongly influenced by the historical approach to economics in general, and to the way in which individual subjects are treated. It is interesting that Palgrave's preface to the first edition of Volume I treats the historical and deductive schools equally: 'The development of the historical school has opened out new and fertile fields, while the wants of those who follow the mathematical method of study have also to be considered'. And in the important entry on the 'method of political economy', induction and deduction are treated as being equally important pillars of positive economics, while it is concluded that 'each method may learn from the results of the other'.[48] Also in the first edition there are a large number of historical subjects,[49] some of which were on the medieval period,[50] and also most contemporary issues (such as prices, insurance, banking, currency, import duties and so on) were placed in an historical context. The contrast, therefore, between the first edition of 1894–99 (or indeed the second edition of 1923–26) and the third edition of 1987, could not be more stark. History and historical method which permeate the first edition are almost totally absent less than a hundred years later. There can be no doubt that the economics profession currently does not see a very important role for economic history.

Despite this neglect by the economics profession of applied work in general and economic history in particular, there is an important place for these activities. As Stigler has suggested, empirical work can and should provide a testing ground for deductive theories, the results of which may lead to the abandonment of some theories and the effective modification of others. This is, in fact, a restricted form of induction –

restricted by the parameters of the deductive theory that stimulated the empirical work. What I mean is that the applied economist tends to see reality through the eyes of the theorist, and seeks out only those variables suggested by the theoretical models. Is it surprising, therefore, that to the economist, both theoretical and applied, the world takes on the appearance of the profession's abstract models? The problem is sometimes compounded by what I will call economists' historians, who also adopt the tunnel vision of economists when dealing with the past and who deal in 'stylised facts' that look very similar to the backward predictions of simple economic models.[51] In these circumstances theory is not just a tool, however much the more perceptive economists may protest, but is a surrogate 'description' of 'reality'. This is precisely where analytical economic history can play an invaluable role. By reconstructing reality, the economic historian can demonstrate the limitations of this narrow focus, and could possibly assist in the revision of theory.

Swimming against this tide of unrealistic abstraction was the emergence of development economics after the Second World War and the consequent collapse of colonialism in Asia and, later, Africa. The critical issue facing newly independent Third-World nations was how to achieve rapid and sustained economic development and growth. For strategic reasons, First-World countries were also vitally interested in this question. Not surprisingly, as this was a dynamic real-world issue of considerable complexity, the existing simple static economic theory was found wanting. Some economists made an eclectic attempt to marshal existing economic theory that dealt with growth in the works of Adam Smith, Thomas Malthus, J. S. Mill, Karl Marx, Joseph Schumpeter, Roy Harrod, Evsey Domar and Alvin Hansen, together with partial theories concerning population growth, technological change, international trade, development strategies, foreign investment and aid, taxation, savings, and the role of government.[52] But it provided little insight or assistance concerning this complex real-world problem. Accordingly, orthodox economics lost interest in this issue, and economic development programmes either disappeared from the subject offerings of economics departments, or were left in the hands of those economists who were either emotionally committed to the issue or who did not have the technical skills to make a career out of theory or econometrics, or it was left to economic historians (such as W. W. Rostow[53]) who became interested in applying their craft to important real-world problems. The highfliers in economics, however, turned this defeat into a victory of sorts by subsequently placing development economics in the same category as economic history – the 'soft' social sciences – and hence not worthy of serious consideration. But the real-world problem, unlike mainstream economists, refused to go away. Development economics, therefore, remains as a rather frustrating branch of applied economics that was largely laid out from the 1950s to the 1970s, although from time to time it attracts a brief round of interest from economists who think they may have discovered the Holy Grail.

Despite the embarrassing failure of growth and development theorists, the mid- to late 1960s saw the high tide of the economics profession in terms of its status, influence, and relative size in the academies. Economists claimed, and were given, much of the credit for the prosperity of the 1950s and 1960s – a period that has since become known in Western economies as the 'golden age'. In this period of unparalleled prosperity, economists felt they had the theoretical key to unlock all the mysteries of the present and future. As they had no need to explore reality, they had no need of history. Many forgot that economics was only a way of thinking, a method for solving problems, a tool box for understanding reality. Accordingly, the focus of the profession in this period was not on making theory more effective as a tool of analysis, but on making it more precise. This was the high point of the arrogance of the economics profession. But the economic circumstances that made it possible were not to last, as history – had economists bothered to consult it – could have told them.

With the emergence in Western economies, after 1973, of major economic problems – slower, even negative, rates of growth, and high rates of both unemployment and inflation – the status and influence of orthodox economists declined. This made it possible for dissenting members of the profession to make themselves heard. Some, like Milton Friedman, revived neoclassical theory (such as the quantity theory) and reshaped it to fit the circumstances of modern Western economies. Others, such as the 'new classical macroeconomists', attempted to resolve the conflict in the neoclassical synthesis by assuming, unrealistically, that markets were highly competitive and that they cleared instantly. In doing so they also revived an earlier attitude to the intervention of governments in economic activity. Hence the outcome of our era of greater economic uncertainty is a greater diversity of analytical and policy views than has been experienced since the Great Depression. All this must be confusing to the politician and businessman seeking economic guidance, to say nothing of the educated lay person. What remains unchanged, however, is the major emphasis placed by the profession on highly abstract, mathematical, and ahistorical models. Even with the obvious failure of economic policy in the late 1980s to prevent deteriorating economic conditions, the need to take real time into account has not been seriously considered.

The declining status of applied economics – the triumph of gameplayer over realist

The history of economic thought is dominated, therefore, by a shift in methodological focus from induction to deduction. In the process, the status of applied economist and economic historian has progressively declined, while that of the theorist has increased. As a result, according to the conventional wisdom, 'the best students were encouraged to concentrate their research efforts on abstract and theoretical, rather than practical and empirical, problems'.[54] While I agree that students were encouraged to focus upon theoretical rather than applied topics, the notion that these were the 'best' students requires further examination. No doubt they were the 'best' at theory, but were they the 'best' at understanding reality? I will suggest in this section that the changing focus in methodology encouraged a new entrant into the profession of economics: the 'gameplayer'; and that this new entrant eventually displaced, at least from the 'top' of the academic discipline, the former inhabitant, the 'realist'. While a healthy discipline in the social sciences requires a balance between these two extreme types of thinker, the economics profession, especially in England, has come to be dominated by the 'gameplayer': applied economics has been left to those who are not regarded as being of the first rank; and economic history has been abandoned entirely to those who clearly have no head for theory at all (such as Cunningham, Ashley, Clapham and their followers). This is the pervading myth of economics.

The gameplayer is a relatively latecomer to the profession. As we have seen, economics was created by realists in the seventeenth century, realists such as William Petty, Charles Davenant, and Gregory King, who were concerned to understand their economy and to resolve urgent problems. The realists held centre stage until the early nineteenth century, when Ricardo's highly abstract and simplified economic models moved the realists noticeably towards the wings and gave the profession a premonition of things to come. The attack on the realists by the neoclassicists began in earnest in the closing decades of the nineteenth century, and although Marshall attempted to play down the hostilities, it was clear by the turn of the century that the gameplayers were dominating centre stage and that the realists had been forced to the fringes of the drama. From then on, pure theory was equated with intellectual excellence and, among potential entrants, the gameplayers – who became known as the 'best' students – were encouraged at the expense of those who had a preference for, and were skilled at, exploring reality. The only challenge to the gameplayers came, ironically enough considering the non-empirical nature of his work, with the publication of Keynes' *General Theory* in the mid-1930s. This seemed to herald a revival in applied economics but, in fact, much of this was due to the upsurge in econometrics that also began at that time. The Keynesian focus on simple macroeconomic variables, however, provided a convenient basis for the new econometric work. The response of the gameplayers was not to concede centre stage to these new players, but to marginalise their influence by giving them their piece of the action to one side of the main show, through separate departments (or subgroups in existing departments) of econometrics. Untainted by these highly technical empiricists, the gameplayers have been able to continue their austere play up to the present.

The gameplayer

To clarify the distinction between the extreme personality types that I have called 'gameplayer' and 'realist', I shall outline their characteristics in greater detail. In pure form, the gameplayer is a person who views life as a series of manageable problems – rather than a complex and possibly unknowable reality – that may be difficult in a technical sense but which are capable of solution. The best gameplayers enjoy, and are good at, working out the connections between variables in technical games and models, and feel uncomfortable when attempting to see patterns in reality through observation. Nobel Prize Laureate Robert M. Solow quite candidly says: 'Most of the economists I know, including myself, don't have any talent for direct observation, nor, by the way, do we have any methods for dealing with it, which is not a reason not to deal with it'.[55] They are good at mathematical exercises not only because they possess the necessary skills, but also because they are able to take the limited objectives of mathematical games seriously. Gameplayers tend to show early promise owing to a lack of complexity of mathematical problems, at least in comparison with the complexity of reality which requires considerable experience and maturity of thought, as well as 'intelligence', to resolve.

The talents of gameplayers are obvious. At all stages of life they demonstrate an ability to solve mathematical and logical problems, and to play 'intellectual' games. At school they are regarded as 'bright', and at university they are seen as the 'best' students. This type of personality also receives more general recognition than other intellectual types from the public. Although the general population may not excel at gameplaying, it is something they enjoy (crosswords, quizzes, computer games, chess and so on) and they can recognise and appreciate the abilities required to excel at it. To the public, the talented gameplayer is the archetypal intellectual. It is not surprising, then, that the gameplayer receives due recognition for his/her skills.

The realist

The realist, in pure form, views life as a complex mystery which he/she attempts to understand, but will not be unduly perturbed if this proves impossible. Indeed, it is precisely the intellectual difficulties and risk of failure that attract the best of this personality type. Intellectual games, however, are not of interest to the realist because they appear to have little purpose in themselves, or because they tell them nothing directly about reality. Reinforcing their set of preferences is their lack of comparative advantage in playing games. Their comparative advantage is in discerning complex relationships in reality or, in other words, discovering order in the chaos of the real world. In contrast to their opposite number, the realists generally take longer to reach their full potential, largely because of the experience and maturity required to do so. And it is the continued accumulation of experience that enables them to produce their best work beyond middle age. Recognition of their powers comes more slowly, if at all, not only for this reason but also because there is no popular parallel with which the public can identify. The road of realism is long and hard.

The rise of the gameplayer

How do we account for the growing dominance of the gameplayer in economics, or indeed in social sciences generally? This is a complex matter and involves forces on both the demand and supply sides. On the demand side, as we have seen, the growing sophistication of industrialised economies has generated a need for highly trained economic technicians with an array of theoretical, mathematical, and technical skills. This has attracted the gameplayer and repelled the realist. On the supply side, the main issues appear to include: technical efficiency; professional practicality; and self interest. Simple abstract models based upon deductive thought provide an easy and effective way to understand the likely relationship between important forces in society. Once expressed in mathematical form, any particular set of relationships can be manipulated with relative ease both in the mind and on the computer. It allows greater and more rapid development, because it is a more efficient way of constructing a body of generalisation about the way reality might work. In contrast, as we have seen, the process of attempting to construct economic principles by induction is far more difficult because, owing to the inherent lack of precision, it is not possible to do so in mathematical form. The precision of deductive thought is both its great strength and its fundamental weakness.

Professional practicality plays a part. As suggested earlier, it is very difficult to reach agreement regarding generalisations drawn from historical observation. Inferences cannot be made with a high degree of precision as they can using deductive methods, and the inferences are likely to change as historical knowledge increases. Deductive methods overcome these problems and increase the probability of constructing a generally accepted set of economic principles. This encourages the gameplayers at the expense of the realists.

Fittingly, self interest plays an important role. Mathematical model building, once the basic techniques have been mastered, is relatively easy, and it can be used to produce academic articles quickly. As one is compelled to assume that academic economists are economically rational, they will adopt those research techniques that maximise their research output, because the 'glittering prizes' – academic jobs, promotion, research funds, awards – go to those with the most publications in the leading journals. How many economic historians, for example, have been awarded the Nobel Prize for economics? In their turn, the successful gameplayers consciously select like-minded students who are called, quite misleadingly, the 'best' students. What is not made clear, or probably even consciously realised, is that what these students are 'best' at is gameplaying. Because academic gameplayers do not have a comparative advantage in the exploration of reality, they have little interest in applied areas in general and economic history in particular. Hence they use their institutional power advantage to discourage and even to exclude the non-gameplayers.

The question then becomes, how has economic history survived at all? Basically, the answer is that most university departments of economics, if not all individual economists, appear to believe that the economic past has something to offer their students, even if it is only moments of light relief or entertainment 'to leaven the burden of real knowledge'. After all, even economic man cannot live by bread alone. Perhaps, like a street directory, economic history helps economists find their way around in the outside world. Or perhaps some employers actually consider some training in economic history to be relevant. For whatever reason, the economics profession agrees that there is a small place for economic historians, provided funds are not in extremely short supply. In Australia over the past three decades the number of economic historians amounted to about 15 to 20 per cent of the number of economists in the old university system.[56] These economic historians have a comparative advantage in exploring reality, and therefore are not directly competitive with economists, otherwise they would get swept into the gameplaying syndrome and economic history would disappear. Only by persuading the clients of economists (or their paymasters) that economic history, as a component in the education of economists, is more important or relevant than previously thought, will it be possible to expand the relative size and perceived importance of economic history.

FROM THE LONGRUN TO THE SHORTRUN

The longrun perspective

For most – possibly about 80 per cent – of the last three centuries, economics has been cast comfortably in a longrun framework. Adam Smith, for example, saw his analysis of the late-eighteenth-century British economy as part of a longer process of transition of the economic system through the economic stages of hunting, pasturage, farming, and commerce. He was concerned to set the process of change in real rather than analytical time. Marx, of course, adopted a similar approach, but for different purposes. It was Ricardo who took economics out of real time and placed it in analytical time for all time. Yet even Ricardo, despite his emphasis on microeconomics, was concerned with the analysis of longrun economic forces – he emphasised population growth, capital accumulation, and technological change, and was concerned with the dynamics of the growth process.

The two contenders who vied for primacy following the demise of the classical school – the neoclassical and historical schools – had very different approaches to time. Obviously the historical school took real time very seriously and attempted, at least in the Germanic version, to place their analysis within a variant of Smith's historical stages thesis. As their concern was with dynamic *processes* they had little time for analytical time. By way of contrast, the neoclassical school dispensed with real time as Ricardo had done, and in addition they abandoned interest in the dynamics of growth even in an analytical framework. But the neoclassicists did have some sense of macroeconomic equilibrium being achieved in analytical time and even this limited vision of the longrun led to an awareness, at least, that in the real world economic changes, of whatever origin, took time to work themselves out. This, in turn, was a sufficient, although probably not a necessary, reason to take an interest in longer-term changes in the economy. Even Marshall, who more than any other neoclassical economist of his time demonstrated a degree of curiosity about history, was not able to integrate his comparative–static microeconomics into a framework of this type. Of course, there were isolated economists, such as Joseph Schumpeter, who continued the classical interest in the dynamic process of economic change, but they had little or no influence on the development path of the profession, at least until economic development became fashionable after the Second World War.

The second assault on real time: Keynes and the shortrun

The neoclassical dominance continued, as is well known, until the appearance of Keynes' *General Theory* in 1936. At a single stroke Keynes shifted the attention of economists interested in macroeconomic issues from the longrun to the shortrun. By the 1950s and 1960s every economics undergraduate could quote (generally incorrectly) Keynes' dismissive reference to the neoclassical concept of equilibrium as being 'the long run in which we are all dead'. Indeed, Keynes' analysis was static as well as shortrun in nature, as he was little concerned with issues of growth. This was left to Harrod and Domar and later the new neoclassicists, but even their approach to growth was highly abstract and unrealistic. While there was a burst of interest in growth models in the 1950s and 1960s, and again the late 1980s and early 1990s (see Chapter 3), they were not concerned with dynamic processes, but rather with longrun equilibria – and in this sense their models were not even theories of growth.[57] The prevailing focus in orthodox economics was basically Keynesian until the mid-1950s when it shifted to a neoclassical synthesis of pre-Keynesian and Keynesian economics until the mid-1970s. In the 1970s the neoclassical synthesis was in its turn challenged by both the 'new classical economics' with its timeless emphasis, and by the brief flowering of monetarist economics, which did bring with it a longer-run approach. Nevertheless, the Keynesian-inspired emphasis upon the shortrun was an important event in the growing dismissal by orthodox economics of economic history. If economists need only focus upon shortrun analysis, there is little point in even glancing back beyond the past year or so, because economic circumstances will have changed and therefore be irrelevant. Keynes' *General Theory*, together with Ricardo's *Principles*, delivered the two most important blows against the use of real time in economics. No doubt in the absence of these two economic giants, a similar change of direction would have been achieved in the profession by others.

At the same time, and for associated reasons, applied econometrics, which in the inter-war period had been concerned to model the process of economic change in a realistic manner, also became less realistic and more 'artificial' in its approach to the economy. As Mary Morgan says: 'Dynamic [econometric] models sank under the tidal wave of largely static Keynesian models, general equilibrium theories, and Cowles simultaneous equations systems'.[58] The nature of econometrics changed after the 1940s in a manner similar to the way economics had been changing since the advent of neoclassical economics. Morgan concludes that:

> applied econometrics seemed to enter a less creative phase. Data were taken less seriously as a source of ideas and information for econometric models, and the theory-development role of applied econometrics was downgraded relative to the theory-testing role . . . Econometric models came to be regarded as the passive extensions of economic theory into the real world, as the 'statistical complement of pure economics' rather than as representatives of a 'synthetic economics' in which theoretical knowledge and information from the real world are combined.

In the 1930s the mathematical theory and statistical model building in econometrics relied more closely upon reality. The resulting dynamic disequilibrium models, which were more detailed and complex, attempted to explain causal sequences and to show the impact of policy on these processes. By the golden age of the 1950s and 1960s, econometric theory and model building had developed a momentum of their own by casting off their associations with reality. These models focused not upon dynamic disequilibrium but upon static equilibrium, in the same way that general economic theory did.[59] They were no longer concerned with the explanation of real time. And the issues that were examined, together with the policy prescriptions that emerged, were inspired by theoretical rather than empirical considerations. The leading edge of applied economics was, by the 1960s, pursuing the same unbalanced development path as theoretical economics, and for the same reasons. This, of course, was only partly due to the influence of Keynes – the intellectual dynamics were the same here as in economics – but the influence of Keynes certainly accelerated the transformation.

As great as Keynes' contribution was to the advancement of economic theory, he did the profession an immense disservice by greatly accelerating its move away from reality. And his legacy is with us today. Most economists see little point in gaining any empirical knowledge about the economy with which they claim to be concerned. For example, D. C. Colander and A. Klamer, by surveying graduate students in economics in 1987, made the disturbing discovery that only 3 per cent regarded knowledge of the economy as 'very important', and as many as 68 per cent regarded it as 'unimportant'. No doubt an even higher percentage would regard economic history, stretching back over the last hundred or so years, as being even more 'unimportant'. This attitude obviously reflects that of their teachers. Confirmation was provided by a survey undertaken in 1988 by the American Economic Association. When asked to rank the 'relative importance now given to various branches of economics', the response of the three groups surveyed (faculty members, recent PhDs, and undergraduates) was, according to the reporter, that 'Economic theory leads the way followed in order by econometrics, empirical economics, and applications; institutions, and literature are at the bottom of the list'.[60] Although economic history is not mentioned here – an interesting oversight – it was bracketed with institutions in the questionnaire. Even economists who protest that they do see a role for economic history are quick to qualify their expression of support by saying, 'but of course we wouldn't want you to teach anything before the Second World War, and we would prefer if you focused upon the period since deregulation of the banking system in the early 1980s'. To this way of thinking the period since the Second World War is the longrun, and anything before that is the *very* longrun and obviously totally irrelevant to current problems.

The essential point, of course, has been missed. There are three reasons for taking a longrun view. First, as economic change, particularly in terms of institutions and even systems, is accelerating, it is necessary to examine even longer periods of time in the past in order to encompass the magnitude of likely future change. Secondly, we begin to understand how an economy is likely to react to future shocks by examining how it has reacted to past shocks, taking into consideration any structural changes that have occurred in the past and may occur in the future. Thirdly, and more importantly, a very longrun perspective that encompasses millennia rather than centuries can detect and analyse the great waves of economic change that have their origin deep in the past and will have their effects far in the future. It is true that, since the 1960s, the revival of neoclassical monetary theory with its emphasis on longrun equilibrium has encouraged a marginal interest in longer-term analysis. Added to this is the interest generated by the neoclassical concern with technological change in dynamic models of growth. However, this 'sources of growth' work (E. F. Denison, Z. Griliches and so on) of the late 1960s and early 1970s employed, as will be discussed in Chapter 3, a very restricted *ceteris paribus* economic model which excluded institutional change. At best this was a constrained and unrealistic form of economists' history. As yet no theoretical model in economics has been able to tap the deepest and most important springs of human progress.

Even the emergence of experimental economics, from around 1950, developed in this shortrun, static framework. This new, but minor, branch of the discipline appears to have resulted from the desire of some economists to replicate the procedures of natural science – to set up controlled experimental tests. They argue that existing 'field' data is not of sufficient quality to test the behavioural assumptions of economic theory. This view is, however, symptomatic of the ahistorical and myopic approach of the profession, as it ignores the great body of survey data collected by applied economists and of historical data collected by economic historians. Indeed, one is led to question whether they have any real interest in data when they are able to say: 'In economics, when things appear not to turn out as expected, the quality of the data is more likely to be questioned than the relevance and quality of the abstract reasoning'.[61] If this is a representative attitude, then experimental economics will not take us very far. In any case, while it may cast some contrived light upon some of the behavioural assumptions in modern microeconomics, it cannot cope with the larger macroeconomic issues, and its scope appears limited to shortrun static problems. It suffers, in other words, from the same defects as the theory it attempts to test, and is unlikely to bring economics closer to the forces of change in the real world.

AN UNCERTAIN FUTURE

The current focus in economics is still directed at highly abstract, mathematical, and ahistorical theoretical structures. There has been no noticeable attempt in universities for economists to work more closely with economic historians, or for economic history to become an essential part of the education of economists. Even in the face of overwhelming evidence that economic policy has failed to prevent economic stagnation, inflation, unemployment, and the rapidly mounting pressure of population on total world resources, the need to take historical reality into account has not been seriously considered. The failure of policy is blamed on politicians, rather than upon the profession's rejection of historical reality. It is doubtful that politicians see this problem in the same light and, as a result, the role of economists as premier policy adviser will undoubtedly be seriously challenged. Already we have seen how successful the environmental scientists have been in attracting the attention of governments throughout the Western world. The fate of the economics profession may well depend upon the ability of economists to provide more realistic policy advice, and this can only be achieved if economists place their activities back within the context of real time.

2 Timely Critics

To forewarn us timely, of what might else have been our loss

(John Milton, 1667)

THE WARNING BELLS

The outstanding success story of the social sciences, particularly during the post-Second World War period, has been the rise of the discipline of economics. Not only does economics possess the most comprehensive and mathematically elegant body of theory outside the natural sciences, but also its graduates have come to dominate most public and many private institutions, and have captured the high ground of policy advice given to national governments and international organisations. Economists have become the soothsayers of our age, conjuring up mystical remedies for the economic ills that beset us. And it is a success that has been formally recognised by the Nobel Prize Committee – the only such recognition given to the social sciences.

There are those, however, who question the basis of this success. Economists have been challenged by two outspoken groups – the historicists, who are convinced that the rejection of history by this deductive science is the rejection of economic reality; and the environmentalists who, since the 1960s and particularly the 1980s, claim that the reality rejected by economists is the ecological system of which the economy is only a part, albeit an increasingly destructive part.[1] As will be argued, the link between these two apparently disparate groups is a romantic rejection of rationalism. While the historicist attack ultimately failed, the environmentalists, who are currently mounting a serious challenge, could break the dominating influence of economics. Yet this need not be so. If economics finally heeds the historicist criticism, which stretches back to the late nineteenth century in the Anglo-Saxon world, the resulting increase in realistic content – including an enhanced awareness of the ecological content – should increase its relevance to real-world problems and strengthen its political influence. This is not a trivial academic matter;

rather it is a matter of the greatest importance and urgency. The warning bells, in terms of persuasive public criticism from historicists and environmentalists, have been sounding for some time.

HISTORICIST CRITICISM

Orthodox economics has attracted the criticism of historicists in the Anglo-Saxon world since the early 1870s. This can be thought of as a belated romantic reaction – which had its origins in Germany during the early nineteenth century – against the rationalism of deductive theory. While some of this criticism has been uninformed, much of it has found its mark and has never been responded to satisfactorily by economists. The most militant and effective attacks made by historicists were launched in the period from the early 1870s to the early 1920s, and were responded to rather ineffectively by the leading economists of the day, such as J.S. Mill, Alfred Marshall, Arthur Pigou, and J.M. Keynes. Since that time historicist criticism has been less authoritative and less effective. Indeed, over the last half-century it has gone completely unnoticed by orthodox economists, who have only bothered to contest the attacks of radical economists of various types – Marxists, neo-Ricardians, and post-Keynesians – mainly during the late 1960s and the 1970s. Unfortunately, since the Second World War, historicist criticism has degenerated into a surprisingly bitter and misinformed attack by historians, with little or no training in theory and statistics, on the work of quantitative and analytical economic historians or cliometricians. This is the unfortunate legacy of the historical school.

To illustrate the nature of historicist criticism over the last century or so I will briefly focus attention upon a representative selection of its leading exponents: T.E. Cliffe Leslie, David Syme, J.K. Ingram, William Cunningham, William Ashley, and John Clapham. The purpose is both to highlight important criticisms of economics that have been neglected, and to differentiate my position with respect to theirs, rather than to provide a comprehensive history of ideas.

The early critics – Cliffe Leslie, Syme and Ingram

The first major Anglo-Saxon attack upon deductive economics was launched in the 1870s by two scholars at Dublin University – T.E.

Cliffe Leslie and J.K. Ingram – and by an Australian newspaper-owner-turned-inductive-economist, David Syme.[2] As the background to the English historical school is discussed elsewhere,[3] I wish to focus in this chapter only upon the nature and style of their attack upon Ricardian political economy. While Leslie and Syme did much of the groundwork for this attack, it was Ingram who, with a more effective militant style, brought the claims of the historical school to the serious attention of political economists in England.

Leslie's ideas were formulated in the 1860s, largely under the influence of the historical jurisprudence of Sir Henry Main, and the German historical school of economics. They first emerged in print in 1870,[4] but reached their fullest expression in his 1876 manifesto, an article entitled 'On the philosophical method of political economy'.[5] Leslie rejected the deductive method of Ricardo as 'illusory exposition based upon simplistic primitive assumptions', in favour of the inductive and historical method employed (occasionally) by Adam Smith, because, he claimed, economic generalisation was relative to the 'different states of society' and modern man was motivated by more than 'the desire of wealth'. 'Our whole national economy is a historical structure, and in no other manner to be explained or accounted for'.[6] And he concludes by claiming:

> The abstract and *a priori* method yields no explanation of the laws determining either the nature, the amount, or the distribution of wealth . . . the philosophical method must be historical, and must trace the connexion between the economical and the other phases of national history.

And again, but somewhat overstating his case, that:

> The abstract *a priori* method . . . has almost entirely lost credit in Germany, and has never had undisputed possession of the field in either England or France. It is repudiated by M. de Laveleye [of Belgium], and by some of the most eminent economists in Italy. Malthus and Say, the two most eminent contemporaries of Ricardo, emphatically protested against it. Mr. J.S. Mill's treatise on the 'Principles of Political Economy' often departs from it, and in his later writings he showed an increasing tendency to question its generalizations. Nor did the founders of political economy, either in England or France, intend to separate the laws of the economical world from the general laws of society. Their error lay in the

> assumption of a simple harmonious and beneficent order of nature, in accordance with which human wants and propensities tend to the utmost amount of wealth, happiness, and good.[7]

Leslie's critique of deductive economics received a widespread and sympathetic hearing, even from J. S. Mill, and some believed that the historical approach would become the dominant school of thought in economics. While Leslie was unable to deflect the deductive school from its ascendant path, he did draw attention to the possibility that deductive theory may not be universally applicable in time and space and, even more importantly, that as the present has emerged out of the past it can only be fully comprehended by an analysis of those forces responsible for its rise.

Another influential, but badly neglected, historical economist was the Australian newspaper proprietor David Syme. Syme's interest in economics arose from his work in the colony of Victoria as owner and editor of the influential and progressive newspaper, *The Age*.[8] In actively promoting the economic interests of Victoria in the late 1850s and 1860s he saw the need to protect local industry from British imports by abandoning free trade and adopting protective tariffs. Like political economists in Germany and the USA, he was dismayed by the conflict between his pragmatic policy conclusions and the policy prescriptions of British classical economics, and he attempted to resolve it by rejecting the deductive approach in favour of a set of inductively-based generalisations. In an article published in the *Westminster Review* in 1871, which owed nothing to Leslie's work, which was only beginning to emerge at this time, Syme challenged both the assumption of material 'self-interest' and the value of the '*a priori*' deductive method. He concludes that: 'Treated by the inductive method Political Economy is a science of the highest practical value; treated a priori, it is not a science at all, but only a scientific artifice, a mere theory of human action in one particular direction, and which has not even the merit of being approximately correct'.[9] This article appears to have attracted Leslie's attention, for when Syme published a follow-up book entitled *Outlines of an Industrial Science* in 1876 (the year of Leslie's manifesto), he thanks his 'friend Prof. T.E. Cliffe Leslie for . . . reading over the proof sheets, and for some valuable suggestions'.[10] The main common ground between Syme on the one hand and Leslie and Ingram on the other, appears to be the influence of Auguste Comte rather than the German historical school. Interestingly, this highly original colonial interloper, although known to both

Leslie and Ingram, was ignored by them. This was not the case, however, in either Germany or the USA, where his rightful role was recognised.[11] Also, Syme's *Industrial Science* was read and annotated by Alfred Marshall, who appeared particularly interested in his novel discussion of the income effect in analysing price changes.[12]

Yet in England it was the fiery rhetoric of Ingram, who was less original than either Leslie or Syme, that attracted most attention to the historical school. In 1878, Ingram presented a presidential address to the Economics and Statistics Section of the British Association for the Advancement of Science, in which he made an uncompromising attack upon deductive economics. This address was aimed at 'a reform of economic studies' and at 'launching this great question on the field of general English discussion'. On this significant occasion, Ingram drew attention to the crisis in English political economy which, he claimed, owed much to the heretical deductive method 'introduced' by Ricardo. In full flight, Ingram accused the deductive school of 'error', 'pretension', 'vicious abstraction', 'metaphysical habit of mind', 'speculative imagination', and of generating a body of thought consisting of 'hasty generalisations from insufficient and ordinary premises'. And in concluding his provocative address he stated his position more simply and clearly than Leslie was able to do:

> (1) That the study of the economic phenomena of society ought to be systematically combined with that of the other aspects of social existence; (2) That the excessive tendency to abstraction and to unreal simplifications should be checked; (3) That the *a priori* deductive method should be changed for the historical; and (4) That economic laws and the practical prescriptions founded on those laws should be conceived and expressed in a less absolute form. These are, in my opinion, the great reforms which are required both in the conduct of economic research, and in the exposition of its conclusions.[13]

It was on the matter of deductive method and economic rationality that Ingram was most eloquent. He described the deductive heresy of Ricardo's 'vicious abstraction' and argues:

> Now, in matters of human life especially, we have only to carry abstraction far enough in order to lose all hold on realities, and present things quite other than they in fact are; and, if we use these abstractions in the premises of our reasonings, we shall arrive at

> conclusions, either positively false, or useless for any practical purpose.[14]

Warming to this theme he claimed that economists employing the deductive method 'wander away from any relation to fact, and lose themselves in a region of nebulous metaphysics'; and, he continued, 'It is a characteristic feature of the metaphysical habit of mind . . . to mistake creations of the speculative imagination for objective realities'.[15] The only corrective, Ingram claimed, is the inductive or historical or 'realistic' method, particularly as a basis for economic policy.

Ingram's widely read attack of 1878, which was followed up with his popular *History of Political Economy* (1888), was influential among political economists during the 1880s and 1890s, at least until Marshall succeeded in developing a set of principles to replace those obsolete theories of the classical school. While this radical and militant critique did not displace the deductive method in economics, it did draw attention to a fundamentally important problem that is likely to emerge in any excessively deductive social science – the fact that there is a temptation to regard theoretical conclusions as a substitute for reality and therefore as an adequate and proper basis for formulating economic policy. One hundred years later this fundamental criticism has yet to be taken seriously by economists.

The main influence of the historical school appears to have been on young scholars. One interesting example is W. J. Ashley who, as a student of Arnold Toynbee's at Oxford between 1878 and 1881 and as a tutor during the 1880s, came under the influence of the English and German historical schools of economics. Originally he hoped that the historical school would replace the deductive school. As he wrote in *English Economic History* in the mid-1880s: 'Modern economic theories, therefore, are not universally true; they are true neither for the past, when the conditions they postulate did not exist, nor for the future, when, unless society becomes stationary, the conditions will have changed'. Elaborating on this theme, he says:

> An increasing number – 'the historical school' in the strict sense of the word – hold that it is no longer worth while framing general formulas as to the relations between *individuals* in a given society, like the old 'laws' of rent, wages, profits; and that what they must attempt to discover are the laws of social development – that is to say generalisations as to the stages through which the economic life

> of society has actually moved . . . this will not only give them an insight into the past, but will enable them the better to understand the difficulties of the present.[16]

While it was a theme he was to touch upon in his 1888 inaugural lecture as foundation Professor of Political Economy at Toronto,[17] it is merely a repetition of historical school doctrine, displaying little zeal and certainly no fire. And his work at this stage involved telling a story about the English economy, not deriving inductive generalisations. Even without historical hindsight it would have been obvious to the perceptive onlooker in the late 1880s that the revolution proclaimed by Ingram a decade before was in danger of being snuffed out. In fact, by the time Ashley accepted the Harvard chair in economic history in 1892, he had lost faith in the possible achievements of the historical school, and was content to accept a lesser role for historical studies by channelling its energies into the development of a separate discipline of economic history. Quite unambiguously he said, in his 1892 inaugural lecture[18] (of the historical schools in both Germany and England) that 'They looked for a complete and rapid transformation of economic science' but 'that no such complete transformation had taken place'. But, he argued, the effort had been worthwhile because the historicists had 'altered the mental attitude of theoretical economists' to the universality of their theory and the hypothetical conclusions of their models; and had convinced them that 'economic forces are not the only forces which move men'. Indeed Ashley had convinced himself that there was an 'awakening – or reawakening – in American and English economies of a sacred passion for the observation of real life'. How wrong he was can be seen in Pigou's reaction to Clapham's critique of economics and, as discussed in Chapter 1, the increasingly abstract development of economics during the twentieth century.

But, Ashley argued, the main contribution of the historical school of economics was the creation of economic history as a separate area of study: 'today the leaders of the school are throwing themselves into detailed research, and are feeling their way towards independent historical construction'. In this effort he felt that economic historians should inform themselves about theory – 'six months' steady work will probably suffice' – but that good books could be written without it by closely studying documents which, unlike statistics, contained 'large and unknown stores of economical instruction'. He was convinced that they should get on with this and terminate their attack on deductive economics: 'It is surely time to cry a truce to controversy . . . let us try for the next twenty years to leave one another severely alone, and see what will come of it . . . if we cannot agree, let us be silent.' And, apart from Clapham's surprising outburst in 1922, they were, and are, silent. They have both gone their own ways, and both are poorer as a result. When he returned to England in 1901 as head of the new Faculty of Commerce at Birmingham, his interest even shifted away from economic history and towards institutional economics. Ashley's pragmatic shift of emphasis, and his mild manner, did not provide the stuff of militancy. That role fell to Cunningham.

The militant historicist: William Cunningham

The public, if not the private, confrontation between Cunningham and Marshall at Cambridge is well known. Their struggle for the right to determine both the nature of economics and the education of economists, burst onto the pages of the *Economic Journal* in 1892. Cunningham, who was both student and colleague of Marshall, was an important figure, not only for his empathy with the historical school but also because of the influence that he had upon the development of the English tradition of economic and social history.

Cunningham was critical of orthodox economists, not because they ignored history, but because they dared to think of themselves as instant experts in the subject. Of course, he had in mind the opening two chapters of Marshall's *Principles*, which in the first edition were meant to provide the historical context for the theory. In his provocative *Economic Journal* article, entitled 'The perversion of economic history', Cunningham complains: 'Economists will not leave it [economic history] alone; they do not pursue it seriously, but try to incorporate some of its results into that curious amalgam, the main body of economic tradition; and the result is the perversion of Economic History'.[19] He goes on to claim that: 'There are two forms in which this perversion of history, through the influence of modern conceptions, may be found: Firstly: In the neglect of serious study of facts. Secondly: In the misinterpretation of facts, even by the most painstaking and careful investigators'.[20] According to Cunningham, the leading economists of both Cambridge and Oxford were guilty of perverting history: Marshall of both forms, Thorold Rogers only of the second. It is interesting to notice the change in emphasis between Cunningham in 1892 and Ingram in 1878; Ingram challenged the economists' use of deduction whereas Cunningham disputed their use of history.

Marshall was so disturbed by this attack that he broke his rule of not replying to criticism. This is fascinating. What was it in Cunningham's attack that drew forth such a long and earnest reply from the greatest economist of the age? The most telling thrust, which clearly pierced Marshall's defences then and those of economists now, is that history is used by economists merely to 'illustrate' their theory, never to falsify it. With less than subtle sarcasm Cunningham mocks Marshall thus: 'If we [the economists] already understand the *principles* which explain industrial and commercial affairs, all that we need do is to look to history for illustrations of what we already comprehend quite clearly'; and again: 'If the facts illustrate his doctrine, he is willing to allude to them; if they do not illustrate it, they merely obscure the great truths he has already formulated in so-called Economic Laws'. And to do so 'prevents the economist from finding out the narrow limits within which his generalisations are even approximately true'.[21]

This hit home. Marshall's 'Reply' is feeble and, indeed, damning on this issue, as he plays into Cunningham's hands. It is worth quoting Marshall at length:

> I once purposed to write a treatise on economic history, and for many years collected materials for it. Afterwards I selected such part of these as helped to explain why many of the present conditions and problems of industry are only of recent date, and worked it into the chapters in question. But they took up much more space than could be spared for them. So I recast and compressed them; and in the process they lost, no doubt, some sharpness of outline and particularly of statement. Nothing was retained that was not needed for use later on. These two chapters are a mere introduction; *they have no claim to be a history*: but they were not written without due consideration of those simple and well-known 'actual facts', which Dr Cunningham supposes me to have neglected.[22]

Cunningham had actually been able to manoeuvre the great economist into denying that his 'historical' chapters were historical. So effective was Cunningham's criticism that, in later editions of *Principles*, Marshall relegated these chapters to appendices. Cunningham was, of course, correct. Marshall's non-historical 'historical' chapters bear the same relationship to history as compressed particle board does to real timber. Yet on this issue, Marshall has much in common with those economists today who deal in 'stylised facts' as a basis for 'economists' history': is there little surprise then that this type of 'history' looks very much like scenarios predicted by their simple economic models?[23] This criticism is just as relevant and necessary today as it was a century ago.

Cunningham's second thrust, aimed at Rogers as well as Marshall, actually missed its mark, although many historicists then and now have failed to realise it. Indeed, this misaimed shot, and the following endless echo of it down to the present, undoubtedly did more to 'pervert' economic history than Marshall's two chapters in *Principles* could ever do. But more of that later. The point to be made here is that Cunningham accused Marshall and Rogers (who, according to the reverend gentleman, at least got his facts right)[24] of misinterpreting the facts of history because they looked at the past as economists – because their constant underlying assumption was economic rationality. Cunningham thundered:

> The underlying assumption against which I wish to protest is . . . that the same motives have been at work in all ages, and have produced similar results, and that, therefore, it is possible to formulate economic laws which describe the action of economic causes at all times and in all places.[25]

He proceeds to identify that motive as being encompassed by the concept of 'economic man'. Unfortunately for his argument, Cunningham was not able to make a persuasive case against economic rationality in the past. His attempt to do so in the second part of his article suffers from the same fault that he correctly accused Marshall of committing – of using history selectively to 'illustrate' his belief (for it is no more) that economic rationality, even if it existed in the 1890s, certainly did not do so in the more distant past. He made no attempt to test his conviction comprehensively.

Marshall was quick to sense Cunningham's vulnerability on this issue. Much of his long reply is devoted to asserting his faith in economic man in the past as well as the present, and in underdeveloped as well as developed economies. Marshall argued that custom is more apparent than real in history, just as it is in the present. He claimed that 'custom is more or less plastic'; that 'the desire to do what is best for oneself is seldom wholly absent from human nature. Habit and prescription do indeed prevent many kinds of original individual action from being thought of at all, and they prohibit others: but there always remains a fringe of debatable ground in which a pushing character can assert itself';[26] and that strange customs in underdeveloped economies in the 1890s, just as in past economies, obscure the real motives – being self interest – of human beings. Drawing on his experience of nineteenth-century India (and even rural England) he concludes that 'under a very still surface there may be running many keen little pursuits of private gain'.[27] Marshall's advice to economic historians on this issue is sensible and shows more empirical insight than Cunningham was willing to grant him. He suggested that:

> the economic historian should . . . explore contemporary customs on the spot; and that, aided . . . by modern analysis, he should use his knowledge to guide him in searching the slender records of medieval customs for subterranean channels of change . . . whether the suggestion is a fruitful one, time alone can show.[28]

Here Marshall is far more perceptive than Cunningham. In Chapters 6 and 7 an attempt has been made to quantitatively test for both the existence of economic rationality and the fruits of that rationality – the growth of per capita income – using 'modern analysis' together with the 'slender records of medieval customs' (which are not as slender as Marshall thought). The suggestion, which was only discovered after the work done in Part II, referred to above, appears to have been fruitful. And it supports the intuition of Marshall, not Cunningham.

Unfortunately, English economic and social history followed Cunningham's conviction rather than Marshall's intuition on this issue. They denied (and in some quarters continue to deny) the reality of economic rationality not on empirical but on metaphysical grounds, and hence rejected (and continue to reject) deductive economics which is based upon this principle. They have interpreted Marshall's retreat over his historical chapters as a complete rout on the use of theory in history. On the other hand, economists following Marshall lost confidence in their ability to use history and turned their backs on the discipline. Accordingly, economics and economic history went their separate ways.

The parting shot – J.H. Clapham

J.H. Clapham was probably the last historicist to draw fire from a leading economist of the day. That it was made at all is a little surprising, not just because Clapham was Marshall's protégé but because by this time other leading economic historians appeared to believe that it was best to bury past differences and to get on with the job. As G. Unwin said in 1908, 'the teachers of economic theory and of economic history are learning to work together in a friendly spirit and to recognise the value of each other's labours . . . Controversial methods are, I believe, entirely destructive of the true spirit of science'.[29] Yet even Clapham's parting shot had little of the penetration of Cunningham's criticism.

The title of Clapham's article criticising Arthur Pigou's brand of economics, 'Of empty economic boxes' – which appeared in the *Economic Journal* for 1922 – is more provocative than the content which, in part, was written as a rather drawn-out and narrowly-focused fable. But for the telling of this fable, Clapham's message to his colleague could have been presented in a few pages. The essence of Clapham's criticism of the state of economics immediately after the First World War can briefly be summarised.

First, economic theory had raced too far ahead of its application to 'life' with the result that it lacked any realistic content. 'Analysis,' he said, 'has often outrun verification.' To illustrate this point he takes the case of the 'Laws of Return' and claims that they 'have never been attached to specific industries; that the boxes are, in fact, empty'.[30] Secondly, Clapham claims that economic theory – although he continues to focus only upon returns to scale – is of limited use in exploring reality. He says, for example: 'Nor can he [the economist] think of an industrial monograph in which profitable use was made of the Laws of Returns in commenting on the things of life';[31] and he was not confident that such a monograph would ever be written. The economist, he claims throughout the article, knows little of reality: 'The puzzled economist has no idea [about reality], and no notion how he shall begin to form an idea'; 'He must, if honest, admit ignorance'; 'he must balance all these uncertainties and ignorances'; and he is forced to 'hazard a guess' about any real-world issue.

Thirdly, Clapham is highly sceptical that theoretical constructs could ever be measured with any precision. He claims, for example, that:

> Constant returns . . . must always remain a mathematical point, their box an empty one. It is inconceivable that a method can ever be devised for so measuring these real but infinitely subtle and imponderable tendencies towards diminishing and increasing returns that someone will be able to say, Lo, here a perfect balance.

> If this is so, constant returns industries may be relegated finally to the limbo of the categories, in company for the present with such still disembodied phantoms as the 'commodity whose elasticity of demand is unity'.[32]

And even if by some miracle this could be achieved, he would remain agnostic about the point of doing so: 'What would be the utility of the knowledge, apart from the satisfaction of a legitimate scientific curiosity?'[33] Of course, Clapham could not be expected to foresee the development in econometrics that occurred from about the time he wrote these sentences, a development that has provided precise estimates of returns to scale and elasticities of demand for real-world industries and commodities. Indeed, in Chapter 6, estimates of the returns to scale on manors in eleventh-century England have not only been presented, but are also used to explain the observed wide variation in manorial size, with no tendency to a single optimal size. As with Cunningham on the issue of economic rationality, Clapham has to concede clarity of perception to the economists.

Finally, Clapham did not hold out much hope for orthodox economics in providing useful policy prescriptions, particularly with regard to Marshall's tax-subsidy proposals. He claimed that the 'impossibility' of separating out the scale effects from those of innovation, prevented the identification of the nature of returns to scale for individual industries. Marshall's tax proposals depended upon this identification. He concluded that:

> Unless we have a good prospect in the near future of filling the boxes reasonably full, there is, I hold, grave danger to an essentially practical science such as Economics in the elaboration of hypothetical conclusions about, say, human welfare and taxes in relation to industries which cannot be specified.[34]

In this claim he was on much firmer ground, for two reasons. For his period, it was impossible to sort out the scale effects from those of technological change, and as Mark Blaug has said: 'This criticism was a crushing blow to the applicability of Marshall's tax subsidy proposals'.[35] Today, econometric techniques can be employed to measure these effects separately, as is done in Chapter 6. Also Clapham is on firmer ground because the attempt, then as now, by economists to base policy prescriptions on theoretical models that have not been applied to the real world, is a highly risky procedure; risky that is for those who wish to implement policy and those who stand to suffer if the interventions misfire. Indeed, it is an irresponsible and dangerous way to proceed.

The contrast in tone between Pigou's reply to Clapham, and Marshall's reply to Cunningham is instructive. Marshall's reply was serious and earnest, Pigou's merely flippant and supercilious. While this difference probably reflects the different personalities of the two economists, it also reflects the different intellectual environment between 1892 and 1922. In 1892 even an economist of Marshall's standing felt the need to treat the claims of historicists seriously; but thirty years later intellectual fashions had changed and history was no longer held in such generally high esteem, particularly by academic economists who could barely remember the time when economics had been taught under the auspices of history at 'Oxbridge'. Deductive economists, after a generation of independence, were supremely confident of their attainments and felt somewhat superior to those who bothered to get their hands dirty with reality's facts. By this time, as discussed in Chapter 1, those who dealt with pure theory were regarded as possessing the 'best' minds, while those who dealt with facts were considered to be intellectually inferior. Marshall appears to have regarded the great historians of his time as equals, whereas one has the impression that Pigou thought they were beneath him.

Arrogantly, Pigou accuses Clapham of 'blurred' thinking, of a 'woolliness in his critique [that] makes the task of reply a little embarrassing'. Pigou, however, should have been embarrassed for other reasons. His distinction between empty boxes and full but useless boxes is not well argued. Surely Clapham was correct in arguing that practical usefulness was a function of realistic application. Also, as Blaug points out, Clapham detected the critical problem with Marshall's tax-subsidy scheme: 'The result was a debate that raged on in the pages of the *Economic Journal* for the rest of the decade, culminating 10 years later in Joan Robinson's *Economics of Imperfect Competition*'.[36] Not bad for a 'blurred' thinker!

After attempting to belittle Clapham in this way, Pigou makes an effort to parry his opponent's criticism that economic theory is unrealistic and impractical. First, Pigou argues that human knowledge comprises the results both of deductive reasoning and of empirical investigation, and that some of the work in both areas could be regarded as 'totally irrelevant to practice'. What Pigou (and many economists since) fail to understand is that all historical research, even of the remote past, informs us about man in society and, as such, can never be irrelevant to practical problems, even when it is trivial. By contrast, the results of some deductive reasoning may be both trivial and irrelevant, informing us not about man in society but merely about the intellectual predilections of the theorist. If the trend in economics – the only justification for which is that it assists us to understand economic reality – is towards impractical theorising, then criticism of the type made by Clapham is justified.[37] Pigou's defence, based on the claim that history is as useless as theory, is not very persuasive. The only proper defence is that, on the whole, economic theory is both realistic and useful. Pigou appears to realise this because the bulk of his reply is concerned to show how returns to scale are relevant to the wider theory of value and to practical policy issues such as the effect of taxes on prices. But despite his confident style, Pigou was unable to handle Clapham's astute criticism of Marshall's tax-subsidy scheme.

Pigou ends his reply by claiming, not very convincingly, that he hoped the future would produce economists who could combine both theoretical and empirical research, in the way that W. S. Jevons had done. He concludes:

> We must endeavour to train up more men of the calibre of Jevons, who are equally at home in both fields. Till we can accomplish that, the next best thing, for those lesser persons who are moderately qualified for the one sort of inquiry and for the other, is to work together in combination, and not to waste time in quarrelling, perhaps on the basis of an imperfect understanding, with the deficiencies of one another's methods.[38]

While a development of this type is essential for the future health of the profession, coming from Pigou it has a hollow ring. There is no evidence that Pigou or his colleagues at Cambridge made any attempt either to train their students in both theory and history, or to promote active co-operation in research with economic historians. It is probably fair to say that by the 1920s the time of the all-rounder like Jevons had passed, but co-operation with economic historians such as Clapham in the more balanced education of economists and in joint research would have prevented the emergence of a receding reality in economics that has brought the profession to a point of crisis.

Clapham's response to Pigou was both modest and dignified. No doubt he could have scored points from Pigou's inadequate response, but he contented himself with the knowledge that at least the economists had noted his complaint against economic theory. Clapham reaffirmed the validity of criticising economic theory for its lack of realistic content; agreed with Pigou that theory is useful in exposing illogical arguments, but, at the same time, was 'very anxious that economic science should be able to do more, and that, where and in so far as it is at present unable to do more, it should make the fact quite clear';[39] and was aware that Pigou had not been able to resolve the difficulty of sorting out the effects of scale and technological change. Finally, he was not impressed by the promise of Jevons' second-coming: 'I had anticipated that the facts and statistics demanded might be, by common consent, at present unprocurable; but I had hoped they might be specified. And now I am paid with a cheque drawn on the bank of an unborn Jevons. Can no one give us more current coin?'[40] Unfortunately no one has.

After the heat and embarrassment of this exchange had dissipated, there emerged a 'comfortable' indifference between economics and economic history, at Cambridge and elsewhere, with each party agreeing to go its own way. As Clapham remarked in his 1929 inaugural lecture:

> Here in Cambridge, I think I may say, economist and economic historian are at peace. We know our limitations. We can sit happily side by side under Adam Smith's great umbrella . . . The Professor of Political Economy will not cry out because I do not read a mathematical article . . . I shall not resent his indifference to [say] . . . the medieval open-field . . .[41]

By the end of the 1920s, therefore, Clapham was willing to concede that economic history was not able to fill the empty boxes with the precision required by the smallish proportion of economists who actually thought they should be filled; he was willing to admit that economic history had to deal with a quality of data that would not satisfy the applied economist. But this was selling economic history short: instead of attempting to be more rigorous, more analytically and quantitatively precise, he decided to accept intellectual second fiddle to the economists and to pursue economic history for its own sake, which, ironically, was the thrust of his criticism of Pigou. As I intend taking up the issue of greater analytical and quantitative precision in Part II, it is worth quoting Clapham further. In discussing the data limitations of the distant past he complains about the difficulty of obtaining *any* firm data, and even:

> If by good fortune the economic historian has one set of measured facts from a remote date – certain truths so far as they go – he will often lack that second set which leads to the illuminating conclusion. He may know the amount of wool exported in 1273 but not its price schedule, *the amount of Danegeld but not the taxable capacity of England*.[42]

It just so happens that the data are available not only for the amount of danegeld, but also for the taxable capacity of England at the time. By using these data it is possible to enquire, as Marshall had encouraged Cunningham and presumably his student Clapham to do, into the motivation of the medieval state. It is curious that Clapham should overlook this possibility, as in the next breath he refers to the 'complete' national record of Domesday Book. It is as if his earlier battle(s) with the economists at Cambridge had broken his spirit and, as it turned out, the spirit of the economic history discipline. If economic history was fit neither to create the boxes of economics, nor to fill them with hard-edged realism, what was its role? Clapham's disappointing answer was that it should become a service activity for both economics and history; in his words, a 'help-study' or, even worse, the economic historian should act 'as a sort of osteologist, a collector and student of historical bones'. What an unfortunate metaphor: the economic historian as the bone-merchant of the social sciences! It was, in Clapham's mind (in 1929), up to economists and historians to put the flesh on these bones: 'I am no rival. Yours is the higher work. I want to help.'[43] Is there little wonder that economists refused to take economic history seriously, when leading economic historians had ceased, by the 1920s, to take it seriously themselves. What would Cliffe Leslie, Ingram, and Cunningham have thought of the sorry state to which their endeavours had led?

It is not surprising, following this admission of defeat by the first incumbent of the Cambridge Chair of Economic History, that many leading British economic historians after Clapham had no interest in interacting with economics, and that they turned increasingly to social history. The creative tension between the theorists and the empiricists, which is essential to the health of the social sciences, was broken. Later professors of economic history did what was expected of them by briefly referring to the usefulness of economic theory (in which few were well-trained) in framing questions for research, although there is not much evidence of its use in their work. There was, however, greater enthusiasm in their expressions of affinity with sociology (E. Power at the LSE; M. M. Postan at Cambridge; R. H. Tawney at the LSE), their rejection of quantitative work (see even A. J. Youngson at Edinburgh as late as 1959); and their desire for greater solidarity with history than with economics (G. Unwin at Edinburgh; W. K. Hancock at Oxford; J. D. Chambers at Nottingham).[44] Economic historians no longer publicly challenged economists for a lack of realism, probably because they neither knew what economists were doing, nor cared. Rather, until the 1960s, they seemed content to work in areas left vacant by economists and other social scientists: as Postan said in his inaugural lecture, 'those fields which the economists are obviously unable or unwilling to cultivate belong to other people'. This was formalised in Britain by creating departments of economic *and* social history. It was not until the emergence in the USA of cliometrics during the 1960s that British economic and social historians showed any fight, not with the economists but with the historians who used the tools of the economist. This was attempted mainly by closing ranks against these intruders on to fields that the old economic historians thought were not of interest to anyone else, and by writing scathing reviews of the new analytical and econometric techniques they employed. This belated rearguard action by the descendants of the romantic reaction against reason has been relatively successful in Britain, as only in the late 1980s did the quantitative economic historians begin to be appointed to chairs of economic history, and then only in departments of economics. For their part, the economists showed mild interest in these new economic historians – they could at least talk the same language – and although their interest has assisted the quantitative economic historians in gaining a few places on the professoriate, they have not let these empiricists deflect them from their chosen path of abstraction.

Historicist criticism, which for five decades had caused leading economists in the Anglo-Saxon world to think hard about the value of their endeavour, had run its course by the early 1920s. Nevertheless, during this period it raised issues of the utmost importance that have never been countered by economists: issues concerning the lack of reality and speculative nature of deductive economics; the risks involved in drawing policy prescriptions from theory that has not been applied to reality; the neglect of historical processes that are involved in the present and will be involved in the future; the misuse of history in the attempt to give theory the appearance of universality; and the limited application of dynamic theory in time and space. It has been possible until recently for economists to neglect these issues, because their role as premier policy adviser to governments has not been challenged. That has now changed. The environmentalists, who are riding a potentially huge wave of popular support, are currently challenging economists for supremacy to influence the future of the planet. It is time that the historicist challenge was taken seriously.

THE NEW 'DISMAL SCIENTISTS'

For about four decades after the collapse of historicist criticism, deductive economics encountered few natural enemies. During this time it developed rapidly and without interference. There was no other academic discipline that was able to challenge the premier policy-advising role assumed by the economics profession. Indeed, only since the late 1960s have intruders from beyond the boundaries of the social sciences dared, or bothered, to enter the economist's domain. The first sign of challenge, which was not seen as a major threat by the economics profession, was the publication in 1968 of a small book with a big impact written by a young biologist, Paul Ehrlich.[45] The author of *The Population Bomb,* who believed intensely that current population trends would overwhelm the world in a further decade or so, attempted to mobilise popular support in order to convince governments to adopt restrictive population policies. The changes in human behaviour that were needed to achieve zero population growth and rectify damage to the environment could only be carried out by decisive government intervention and control. While Ehrlich was highly successful in stimulating public debate about the issue of zero population growth – his book was reprinted about twenty times in the first two years after publication – economists and governments resisted implementation of his dramatic proposals, because they were convinced neither about the inevitability of the dismal outcome that he predicted – after all, Malthus (who was not quoted by Ehrlich) had made the same prediction some two hundred years previously – nor about the desirability of intervening so heavily in democratic societies. But since then there has been an amazing upsurge in public emotional support, particularly among the young, for environmental issues. It is fascinating to realise that this popular wave of environmentalism upon which the ecologists and their fellow travellers are riding derives its energy from an intuitive form of romanticism that can be compared, and contrasted with, the more intellectually inclined romanticism of the early nineteenth century that influenced the historicist critics of economics. But this is the only link between the timely critics.

The Population Bomb was quickly followed by *The Limits to Growth*, published in 1972 by the Club of Rome, an international group of scientists and industrialists brought together by Aurelio Peccei to analyse 'the present and future predicament of man'. Interestingly, their first meeting occurred the month before Ehrlich's book was published. *The Limits to Growth*, which was the Club of Rome's first and most famous book, attempted, on the basis of a computer model developed by systems engineers at MIT, to predict the long-term outcome of current trends in population, resource exploitation, food production, industrialisation, and pollution. They summarised their conclusions as follows:

1. If the present growth trends in world population, industrialization, pollution, food production, and resource depletion continue unchanged, the limits to growth on this planet will be reached sometime within the next one hundred years. The most probable result will be a rather sudden and uncontrollable decline in both population and industrial capacity.
2. It is possible to alter these growth trends and to establish a condition of ecological and economic stability that is sustainable far into the future. The state of global equilibrium could be designed so that the basic material needs of each person on earth are satisfied and each person has an equal opportunity to realise his individual human potential.
3. If the world's people decide to strive for this second outcome rather than the first, the sooner they begin working to attain it, the greater will be their chances of success.[46]

Only by international co-operation could this 'overwhelming' problem – the 'transition from growth to global equilibrium' – be tackled with any hope of success.

While the model employed by the Club of Rome was based on highly simplified assumptions, it was far more sophisticated than the largely intuitive approach of Paul Ehrlich, and as such was more of a challenge to the work of economists. Also, while the Club of Rome recommended immediate action, their prediction of doom stretched out over the following century rather than the following decade, as Ehrlich had suggested, and they were more cautious about advocating strong government intervention. While *Limits* attracted much excited attention in the early 1970s, the obvious lack of reality of its assumptions, the closing of ranks by the economics profession, together with the unexpected collapse of the 'golden age' of rapid growth, were sufficient to halt this threatening intrusion of outsiders on to the territory of economists. The preoccupation of governments throughout the Western world after 1973 was not how to achieve longrun global equilibrium, but how to overcome the unexpectedly sudden shortrun problem of growth slowdown and the emergence of high rates of unemployment and inflation. National governments dropped any limited interest they may have had in the Club of Rome and turned for support to their old advisers, orthodox economists.[47]

But the collapse of high rates of economic growth has not proved to be an effective barrier against further and more determined attacks by natural scientists, who have been aided over the past decade by increasing popular concern with the state of the environment. While the emergence of this concern for the environment has complex causes – with its roots in the romantic search in the late 1960s and early 1970s by the baby-boomers for an alternative to growthmania and the prevailing urban culture – the frustration at the performance of the economy since the mid-1970s may have led those who had been adversely affected to seek satisfaction elsewhere. With this growing concern about the depletion of the world's natural resources and the pollution of its environment have come new attacks – from biologists, ecologists, climatologists, and geneticists – on the economy, which is seen as the enemy of the ecology. What began in the late 1960s and early 1970s as a critical discussion of economic growth, has in some instances degenerated into a frustrated attack upon economists (as well as industrialists), as if they were responsible for the 'greed' of human nature that is allegedly pushing mankind to the edge of darkness. Currently the most outspoken of these critics are the evergreen Paul Ehrlich and a new chum, David Suzuki.[48] Both these natural scientists have made a highly successful effort to tap into the rapidly growing popular concern about environmental issues. The environmentalists may not be able to defeat the economists intellectually on their own territory but, unlike the historicists before them, they may well be able to marshall sufficient popular support to outflank their economic rationalist enemy in the political domain.

Mainstream economics has come to be regarded by environmentalists as the main enemy of the planet. While this implacable opposition to economists varies in its expression, from the more sophisticated critique of Paul Ehrlich to the populist denunciation of David Suzuki, it is all aimed at replacing mainstream economics with a new discipline called 'ecological economics', and making the new economists the agents of environmentalist policy – of changing the economic system to accommodate the views of natural scientists. The danger to orthodox economics and to the stability of economy and society, which is based upon massively growing popular support, is quite clear.

Ehrlich's views on economics and economists will briefly be reviewed because of his leading role in the environmental movement. In the late 1960s and early 1970s, Ehrlich appeared to believe that economists could, and probably would, assist in transforming the world before it was too late.[49] Over the following decades, with the apparent growing urgency of environmental problems – including the build-up of greenhouse gases[50] and damage to the ozone layer – and the continued intransigence of economists in the face of territorial incursions from environmentalists, Ehrlich's approach to economics and economists has become more militant. For example, on the recent CBC programme 'A Matter of Survival', Ehrlich claimed that: 'Less than 1 percent of the economists in the world have even the vaguest idea of how the world works, and yet politicians listen to economists'.[51] And again: 'economists are trained in ways that make them utterly clueless about the way the world works. Economists think the world works by magic' as it is a closed system with no interaction with the ecology.[52] Indeed, he accuses them, unlike his fellow natural scientists, of living in a world of make-believe: 'economists, as I've said before, live in a fairy-tale world and believe fairy tales. Scientists know better.'[53]

Why does Ehrlich hold this militant stance regarding economics? His latest book, with Anne Ehrlich, *The Population Explosion*, holds the answer. Ehrlich insists that the only way to prevent the destruction of the planet in the twenty-first century is by defusing the population explosion and reducing the stock of population to sustainable levels. The main obstacle to doing this, according to Ehrlich, is the refusal of economists – who advise governments and businessmen – to abandon their preoccupation with 'growthism'. Economists fascinated with growth are unable to see the urgency of controlling the population explosion. He also acknowledges, somewhat reluctantly, that the co-operation of economists will be required in order to change the economic system to accommodate the policies of environmentalists. It is essential, therefore, to reject mainstream or neoclassical economics, and to replace it with the new discipline of 'ecological economics' which will enshrine the ideas of scientists concerning the natural world.

This will involve nothing less than the complete 're-education' of economists. This economic re-education can be achieved by environmentalists reaching out through formal education and the media to the public, who in turn will put pressure on governments to obtain new policy advisers not obsessed by 'growthism', and who will resist 'growthism' in the market place. It is all a matter of intellectual persuasion to change the nature of economics: it would appear that there is a higher rationality than economic rationality.

In his critique Ehrlich completely rejects material self interest as a driving force in human society. Indeed, the growth orientation of businessmen and politicians is blamed upon the obsession of economists with growth, rather than the economic rationality of economic agents. As Ehrlich claims in *The Population Explosion*:

> It has long been clear to ecologists that the extreme growth orientation of mainstream economics is a major reason that politicians, businessmen, and others advised by economists, as well as the public at large, fail to recognise the increasing seriousness of the population crisis in particular, and the deepening predicament of *Homo sapiens* in general.[54]

Like the historicists, the environmentalists wish to replace economic man with a more malleable creature of their own making, who I will call 'natural' man. Ehrlich provides an outline of 'natural' man when discussing what he calls 'our evolutionary handicap', by arguing that the human mind 'evolved as a tool for keeping us alive and maximizing our reproduction'. But he does not see the 'reproduction maximizing' of natural man as an unalterable drive: modern man, unlike his earlier counterpart, pays 'more or less heed to the "maximize your reproduction" message engraved in their genes – in part because there's much less danger that their children won't survive to reproduce. Cultural evolution clearly can override biological evolution'.[55] In other words, modern 'intellectual' man can override primitive 'natural' man. Intractable economic man has been completely banished. Clearly this is an ahistorical assertion, because it is well known that 'primitive' societies, such as Aboriginal society in Australia before 1788, deliberately controlled their rates of reproduction.[56]

Ehrlich, and other environmentalists, are keen to dismiss longrun growth as 'straightforward nonsense' because they are committed to the view that it is not an essential characteristic of human history. They regard growth, along with economic man who gave rise to it, as a modern invention – occurring only since the Industrial Revolution – which can be dispensed with as soon as the power of the economist is broken. Ehrlich says: 'In fact [but he supplies no facts] the state of the economy over most of the sweep of human history was, by the standards of economists today, stagnation'.[57] As will be shown in Chapter 7, Ehrlich is absolutely wrong, as growth has persisted over the millennia. Of course, if longrun growth has been a persistent characteristic of human history, then growth, together with the underlying driving force, will be difficult to root out. This is why the ecologists wish to assume it away.

Ehrlich has firm views on the nature of mainstream economics and the need to replace it with a new science-oriented economics. As he says: 'the discipline [of economics] focuses too little on important questions of policy and too much on learning to manipulate esoteric mathematical theory based on preposterous assumptions'; and he goes on to say: 'the new interdiscipline of ecological economics . . . should become a major area of specialised education, and replace neoclassical economics as the central focus of economics departments'.[58] The new discipline of 'ecological economics' will be based on the central assumption that 'society's first priority must be to keep nature's house in order', and it would include 'considerable instruction on the basics of how the physical–biological world works'.[59] This attack on mainstream economics is just as fundamental as that of the early historicists, and could be far more lethal if it is able to mobilise popular support. It does highlight, however, the major weakness in economics – its preoccupation with deductive/mathematical models, rather than with reality.

But the attack on economists is only a means to an end. The objective of the environmentalists is not only to 'convert the economic system from one of growthism to one of sustainability', it also involves the 'transformation of society'.[60] Nothing less than an entirely new social and economic order is required to accommodate the new ecological vision. Not only is this transformation necessary, it is also possible that 'social behaviour and economic systems can be modified in a few years'.[61] While they regard the ecology as fragile they appear to believe, unlike empirical social scientists, that social institutions and economic systems are both robust and capable of rapid change. As is quite clear to economic historians, and any observer of recent events (June 1991) in Eastern Europe, economic systems cannot be transformed rapidly without widespread chaos and possibly complete collapse, with a resultant emergence of repressive military dictatorships. Re-education, therefore, should not be limited to economists: ecologists could benefit from some acquaintance with the social sciences.

Less rigorous environmentalists who depart from their specialist field, such as the populist David Suzuki, have only half understood the criticism of Ehrlich, and are responsible for spreading incorrect and silly ideas about economics and economists. While these ideas are easily countered by economists, they are sufficiently outspoken and simplistic to gain wide currency through endless media repetition. A major rhetorical device is the use of science fiction – what the world *will be* like in 2040 as a result of the predictions of natural scientists being ignored – to persuade the media-watching public. In their book *A Matter of Survival*, A. Gordon and David Suzuki take up Ehrlich's theme that economists do not understand the wider reality of the physical world; that they are unfit to 'run' our lives; that they are in part responsible for the 'greed' and 'mindlessness' of our existing economic systems; and that they have committed the world to a suicidal fixation with economic growth. These themes were borrowed from Ehrlich, but in the hands of Gordon and Suzuki they take on a more simplistic, populist, and no doubt more dangerous, form.

Taking up one of Ehrlich's themes, Gordon and Suzuki assert that 'fairy-tale economic thinking has given us a skewed view of the world and has pushed us to the verge of a calamitous future'.[62] To illustrate this 'fairy-tale thinking' they draw attention to the alleged faith of economists in economic growth at any price; to the failure of economic prediction; to the refusal of mainstream economics to analyse the interactions between economy and ecology; to the absurdity of the economists' measure of economic activity, GDP, which counts the cost of pollution and resource destruction as an increase in income; and, most curiously of all, to the alleged belief of economists in an infinite supply of resources. This attack on economics and economists is poorly informed, naïve, largely incorrect, and it is easily refuted. As my intention here is merely to outline the nature of the environmentalist criticism of economics, I will not attempt to refute these charges. I will, however, focus upon their discussion of economic growth and the underlying drive of material self interest, both because these issues are dealt with empirically in Part II of *Economics without Time* and because, ironically, this is the weak point of the environmentalist attack. Both economists and environmentalists are on dangerous ground because of their shared ahistorical approach to current problems of both economy and ecology.

Economic growth has been isolated by environmentalists – because it can be used by political leaders to justify further increases in population – as the basis of their attack on economists and economics. In fact, this attack is misconceived and, moreover, unnecessary. It is misconceived because 'growthism' is not the central concern of neoclassical economic theory. Neoclassical economic theory focuses upon the mechanism and effectiveness of the distribution of resources, goods, and income in the shortrun. As shown in Chapters 1 and 3, modern deductive economics does not concern itself either with real time or even the analytical longrun, and its various attempts to model the growth process since Adam Smith have been sporadic and largely unsuccessful. The core of modern economics is concerned with the analysis of comparative static equilibrium conditions and not with dynamics. It is the historicists, not the economists, who have retained a key interest in growth, not necessarily because they consider it essential to the present and future, but because it has existed in the past. And this criticism by ecologists is unnecessary because 'growthism' has not been imposed from above by economists, but is an outcome of the desires of individuals bubbling up to the level of regional and national political circles which seek economic advice, not on the desirability of growth but on how it can best be achieved. They do so because they perceive it to be important to their electorates. Clearly, radical ecologists do not wish to face this reality, because it makes their task of changing the world even more difficult. To change the nature of economics may be difficult, but to change human nature, as opposed to human behaviour, is impossible.

The inescapable fact, whether ecologists like it or not, is that economic growth has been an important part of human history for a very long time. Even historians are only just starting to realise the full implications of this fact. Human civilisation could not have emerged 5000 years ago in the fertile crescent without the growth of surpluses extracted from society by ruling elites. As will be shown in Chapter 7, high rates of growth that can actually be measured were experienced, at least in Europe, over the past millennium. Gordon and Suzuki, who follow Ehrlich in this matter, have blundered badly when they claim: 'If you look at history you see that growth has been part of our culture for only the past 200 years. Before that material growth was negligible and not really important'.[63] But they do not stop here. They seem to recognise that growth may be the outcome of some instinct imbedded in human society, even if it has only been there since the Industrial Revolution, for they say in the same context: 'What is being asked of us in the name of survival is nothing less than to take a new measure of what it means to be human, to face the future stripped of the cloak of materialism'.[64] And at a later stage they elaborate this need to change human attitudes as follows: 'To renegotiate our contract with nature, to alter our worldview, the party of the first part must be changed from *I* to *we*. The *we* is all humanity. Our future lies in group survival and group success, not in individual achievement.'[65] They appear relatively confident that as economic growth and the underlying driving force of material self interest have existed for only the last 200 years, it should be possible for society to reverse this modern trend, and thereby enable the human race to return to a state of stationary equilibrium.

Unfortunately for the environmental sciences, economic growth and economic man can be shown quantitatively to have existed at least throughout the past millennium. Medieval man was economic man, not moral (or communal) man, and he lived in a relatively dynamic society that enjoyed longrun growth. And there is every reason to believe that this has been the case since the beginning of civilisation, although the beneficiaries of growing surpluses extracted from the peasants were the small, but growing, ruling elites. Part II of this book will provide the quantitative evidence for this claim.

In more speculative vein, it would seem that the dominating characteristic of mankind – material self interest – which generated the drive for our species to rise from obscurity to gain unprecedented control over other species, is also the characteristic that will either resolve our current environmental constraints or lead to destruction of life as we know it. Man may change the world but he cannot change his nature, despite the naïve claims of the ecologists. Any attempt to do so will be repressive and will probably lead to the collapse of human society. There is little likelihood of obtaining a voluntary acceptance in developed countries of the major reduction in living standards that the policies of radical ecologists require. If economists have a faith, it is not that resources are in infinite supply – after all, their discipline is the science of scarcity – but in the boundless ingenuity of the unshackled human mind. But of course, like the ecologists, they may be wrong.[66]

Yet the ecologists are correct about one matter. Economists do not have a sound grasp of the real world, either of its economy or its ecology. History, however, provides a context for, and a perspective upon, both senses of reality. Economic historians have reconstructed the economy, both on its own, as economists prefer to view it, and as part of a wider ecological entity, as demanded by ecologists.[67] While my main aim is to point out the importance of historical reality to economists, one could similarly demonstrate the importance of economic history to the environmentalists. Both deductive economist and environmental scientist employ generically similar computerised deductive models, and both are unaware of the motivation of, and social path taken by, man in his struggle for ascendancy on this planet. It is ironical that ecologists should accuse economists of the very crime against reality of which they themselves are guilty.

IT'S TIME TO TAKE TIME INTO ACCOUNT

The purpose of this book is to say that it's time to take time into account. Economic history, ranging from the present into the remote past of 1000 to 3000 years ago, can not only contribute to economic generalisation but can, with a degree of precision sometimes equal to that of applied economics, give realistic content to economic theory. Given the unbalanced development of the economics profession, which has accelerated since the time of Pigou, it is essential to bring back real time. This will not be achieved by acquiescing in the intellectual superiority of economics, as Clapham did, nor by militantly rejecting deductive economics, as Cunningham and the earlier historical economists did, but by forcefully demonstrating to economists and, more realistically, their clients, that without rigorous economic history their house will fall before the onslaught of the environmentalists. The objective is not to displace deductive economics, but to become an equal partner with it in building a more balanced, realistic, and viable profession of economic science that can take a major part in building a secure future for mankind.

3 The Timeless Approach

In wandering mazes lost

(John Milton, 1667)

GROWTH WITHOUT TIME

Modern economics has been purged of time. But what are the nature and limitations of this timeless approach to economic problems? In discussing these issues I will explore the nature of economic theory, the way in which it is applied to the real world, and how policy prescriptions are formulated by economists, all by focusing upon the central problems – that are taken up in Part II – of economic growth and fluctuations. The theoretical discussion traces the development of growth models from Adam Smith to the 'new' growth theory of the late 1980s and early 1990s. The examination of applied economics is concerned with the emergence of growth accounting in the 1960s; and the policy section focuses upon the changing instruments of policy formation during the last few millennia, together with the way in which economists have related policy to the key issues of growth and fluctuations in more recent times.

THEORY

The history of economics exhibits a continuing, if sporadic, concern for the growth of nations. This is not surprising, as economic change, it will be shown in Chapter 7, has been the normal expectation during the past millennium. While growth has long been of interest to policy-makers, the first systematic treatment of the issue can be found in the works of the classical economists, including Smith, Malthus, and Ricardo. This interest was carried forward by Marx; Schumpeter; Harrod and Domar; the neoclassicists, including Trevor Swan, Robert Solow *et al.*; and more recently by the 'new' growth theorists. Without wanting to prejudge the outcome of the 'new' growth theory, it seems fair to conclude that the most impressive work on growth theory was produced before the First World War. The growth analysis of the classical economists – together with the work of those who relied heavily upon this foundation, including Marx and Schumpeter – was a central feature of their wider analysis of the capitalist economic system, and the insights it contains depends to a considerable degree on historical knowledge. Growth theory since the First World War, however, has been a highly specialised and largely detached area of study undertaken within the broader framework of economic analysis, rather than as an integral part of a grand system, and certainly without reference to historical reality. These more recent theories focus on a very limited number of variables and as such are of little use in explaining real-world, or historical, growth *processes*. What they can do, however, is to analyse a certain *outcome*, such as 'steady state', regular cycles, or convergence with or divergence from an equilibrium growth path. They can achieve this because they have been constructed to do so. As the perceived economic problems have changed, new models that can provide these outcomes have been constructed and then abandoned. There is no general theory that can encompass these *ad hoc* partial theories. No single theory can either explain all past growth outcomes (let alone processes) or successfully predict future growth outcomes.

The classical model of growth

What can the various growth models tell us about reality? Is it really possible to explain a process involving real time by adopting a timeless approach to model building? In order to address these and similar questions, a brief survey of the main types of growth model has been undertaken. The intention is to isolate and discuss the main features of these growth models in a general rather than a mathematical way. There are many good technical surveys elsewhere.[1]

The growth models of the classical school were concerned to examine and explain the dynamic forces underlying the British economic system of the eighteenth century. This interest was stimulated by a desire to ensure, through appropriate policies, that the British population gained the greatest material benefit from these forces, not only for its own sake but also to achieve the greatest general progress of human civilisation.[2] In the generalised classical system – in which there is one sector, agriculture (although this can be expanded to include other sectors), three factors of production (land, labour, and capital), and three corresponding economic groups (landlords, workers, and capitalists) competing for shares of the social surplus – the various elements of production, exchange, distribution, and accumulation are integrated to explain not only how the economy operates at a point in time but how it changes through time.[3]

Central to the process of growth is the accumulation of capital, which is stimulated by that part of the social surplus – profits – that accrues to capital. Only capitalists invest their surplus, because workers expend their wages entirely on the means of subsistence, and landlords devote their rents entirely to unproductive consumption. During a growth phase, the sequence of causation is as follows: profits rise; funds are invested in capital equipment, possibly embodying a degree of technological change (resulting in a further division of labour); the demand schedule for labour (MP_L) shifts to the right; nominal wages rise; population increases; the margin of cultivation is extended; diminishing returns in agriculture are encountered; production costs rise; profits are reduced (rents increase); and the rate of capital accumulation and population growth decline until eventually the economy reaches the stationary state. Figure 3.1 provides a simplified diagramatic explanation of this process. Although the classical economists understood that technological change and trade would delay the tendency towards the stationary state by shifting the demand schedule of labour outward, they believed that the ultimate fate of the mature capitalist economy was stagnation.

The fundamental flaw in this model, which J.S. Mill was supporting as late as the mid-nineteenth century, is that it was based upon assumptions that were not closely related to reality during the Industrial Revolution. As is well known, the classical economists underestimated the role of both technological change and international trade in counteracting the move to a steady-state equilibrium. In part this was because they focused upon a pre-industrial economy, and in part because they did not understand (as many still do not) the dynamics of pre-industrial societies. Had they been better historians they would have realised that the British feudal and medieval economic systems had experienced rapid growth (in per capita terms) involving technological change and international trade, as well as diminishing transaction costs over prolonged periods of time. Technical change overwhelmed any tendency towards diminishing returns in agriculture, and international trade and foreign investment provided expanding opportunities for continued growth.

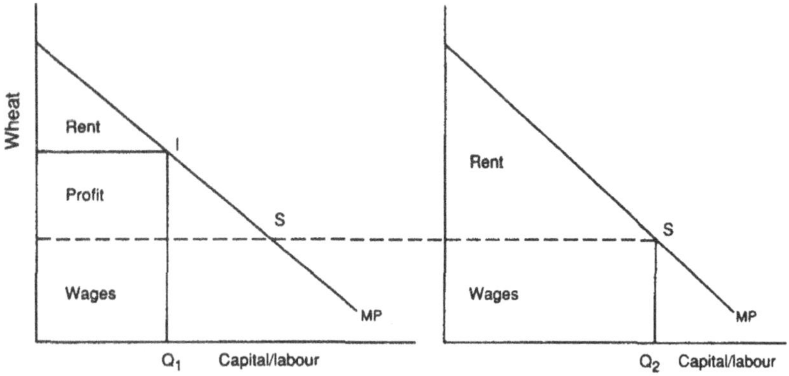

Figure 3.1 The classical model of growth and distribution

Notes:
1. An increase in the combined input capital/labour (in response to profits) from Q_1 towards Q_2 will reduce the marginal product of the variable input. Profits fall and rents rise, while wages remain at subsistence level.
2. This process will continue until the marginal product has fallen to S, the stationary state, where it is equal to the subsistence wage, and profits have been eliminated.
3. The particular interest of this model is that it has traditionally been used to characterise the Middle Ages and to explain the alleged lack of growth.

The timeless approach of the classical economists distorted their view, not only of the future, but also of the past. Their view of the past was obscured by the limitations of their abstract model, particularly concerning their theory of distribution. We have already seen that the classical economists regarded growth as a function of the proportion of the social surplus that found its way into the hands of capitalists, because both landlords and labourers used their shares unproductively. As feudal societies were dominated by powerful landlords, they argued, very little of the social surplus could have been invested productively, and hence feudal societies must have been unenterprising and largely stationary. This conclusion is important not because it was correct, which it certainly was not, but because it became the conventional wisdom in England about pre-modern societies that has been subscribed to by economist and non-economist alike. Indeed, it is still the basis of the explanation employed by some eminent scholars today.[4]

Marx's model of growth was a variation on the classical theme. While Marx accepted the importance of capital accumulation and its relationship to the *rate* of profits (rather than just the *size* of profits, as in the classical model), he saw a greater role for technological change, which he regarded as the main driving force of the capitalist system, and the interaction with the world economy. On the other hand, he rejected the classical relationship between wages and population expansion, and substituted arguments about the 'reserve army of labour' and changing labour force participation rates. Despite these changes to the classical argument, the outcome of Marx's model was much the same. Although capitalists attempt to maintain profits through innovation, the growing capital intensity of production (owing to the labour-saving nature of innovation) reduces the profit rate, places increasing pressure on wage rates, increases unemployment, and finally causes the capitalist system to collapse after experiencing increasingly severe fluctuations. Despite the many insights that Marx's economic system provided, it distorted his view of the past, present, and future. His view of reality was obscured, as it was for the other classical economists, by the inflexible use of theory. In Marx's case this largely involved adherence to the principle of dialectical materialism, which was a philosophical/political, rather than an economic, explanation of social change. He was committed to the idea of destructive class conflict – rather than the classical idea of creative competition – and an economic system that would generate this conflict. It did not seem to occur to him that such a system was economically irrational. Ironically, the centrally-determined system that replaced capitalism in some countries, notably in Eastern Europe, under the inspiration, if not the direction, of Marx was economically irrational and, as a result, finally collapsed at the end of the 1980s.

The Schumpeterian growth model

Schumpeter's growth model, like that of the classicists, was part of his theory of the entire capitalist system. Indeed, his concern was with dynamic processes underlying society as a whole. In his own words:

> Economic development is so far simply the object of economic history, which in turn is merely part of universal history, only separated from the rest for purposes of exposition. Because of this

> fundamental dependence of the economic aspect of things on everything else, it is not possible to explain *economic* change by previous *economic* conditions alone. For the economic state of a people does not emerge simply from the preceding economic conditions, but only from the preceding total situation.[5]

Needless to say, this is in marked contrast to the more narrowly focused growth theory that has emerged since the First World War.

Schumpeter's analysis begins with a discussion of 'the circular flow of economic life' in which all economic activity is repetitive, and undergoing a predictable routine. He focused upon an economic system in longrun general equilibrium, in which all factors are fully employed and which are 'paid' their marginal products, whether in the market or the household sectors. Economic growth only occurs when this longrun equilibrium is disturbed. He explains:

> Development . . . is a distinct phenomenon, entirely foreign to what may be observed in the circular flow or in the tendency towards equilibrium. It is spontaneous and discontinuous change in the channels of the flow, disturbance of equilibrium, which forever alters and displaces the equilibrium state previously existing. Our theory of development is nothing but a treatment of this phenomenon and the processes incident to it.[6]

And it is innovation that causes this 'disturbance': innovation embodied in new plant and equipment. Innovation of product or process in this model generates super-normal profits for the leading entrepreneur, which in turn encourages other entrepreneurs to follow suit. The original innovation may also call forth a series, or 'cluster', of complementary innovations. Accordingly, there is a burst of investment financed by credit expansion, which in turn produces an increase in employment and output. As investment increases, and if there are no further innovations, profit rates fall until they reach normal levels, and new investment ceases. Once more we are in equilibrium.

As Schumpeter explains, the innovation-led boom 'altered the data of the system, upset its equilibrium, and thus started an apparently irregular movement in the economic system which we conceive as a struggle towards a new equilibrium position'.[7] In this model the progression from one equilibrium position to another passes through a cyclical movement of boom and depression. Schumpeter argues: 'the boom . . . creates out of itself an *objective situation*, which, even neglecting all accessory and fortuitous elements, makes an end of the boom, leads *easily* to a crisis, *necessarily* to a depression, and hence to a temporary position of relative steadiness and absence of development.'[8] The boom involves a disequilibrium in which there is 'overproduction', 'skewness', 'the appearance of disproportionality . . . between quantities and prices of goods' and between new and old businesses and industries, and also speculation.[9] The depression is an inevitable outcome of this disequilibrium and 'the driving impulse of the process of depression cannot theoretically stop until it has done its work, has really brought about the equilibrium position . . . Nor will this process be interrupted by a new boom before it has done its work in this sense' owing to the uncertainties about 'new data' which make the 'calculation of new combinations impossible'.[10]

Schumpeter's theoretical discussion of the growth process is clearly informed by observation of historical processes of change. In fact, it is probably the most successful and influential attempt in the history of economic analysis to use an inductive–deductive approach. This was due in part to Schumpeter's training, which was both theoretical (without being mathematical) and empirical, and in part to his genius for observing and interpreting real economic processes. He was a realist rather than a gameplayer. Accordingly, it is to Schumpeter rather than any other growth theorist that those who wish to reconstruct real-world growth processes turn – as will be shown in Chapter 4.

Schumpeter's overall grasp of reality was superior to that of the classical economists and Marx. He was able to distinguish between the capitalist and the entrepreneur as well as between invention and innovation; he saw the connection between innovation and the rate of profit; he could see how technological change could prevent the emergence of the steady state; and he was able to integrate a theory of cyclical fluctuations with a theory of growth. His model does not just provide a method for thinking about causal relationships, it provides an unparalleled basis for understanding and reconstructing real, or historical, processes of change.

Keynesian and neoclassical growth models

Modern growth theory, which has emerged largely since the Second World War, can be distinguished from earlier work by its more abstract, ahistorical, and disembodied nature. The initial work by

R. F. Harrod (1939) and E. Domar (1946) was inspired by Keynes' *General Theory* (1936).[11] As is well known, the *General Theory* presented a macroeconomic model that attempted to show how it was possible for short-term equilibrium to occur at less than the full-employment level of national income. As Keynes' analysis was comparatively static in nature, Harrod attempted to translate this system into dynamic terms. But in fact we are not given any insight concerning dynamic *processes* as we are in Schumpeter's work. To Harrod, dynamic analysis meant the analysis of secular changes in output, capital accumulation, employment, and inflation. While he recognised technology, along with these variables, as an important determinant of growth, and as capable of affecting the ratio of required capital to output (depending on whether it is 'neutral', 'capital-saving', or 'labour-saving') it is not treated endogenously. This model is rather one-dimensional in comparison with Schumpeter's analysis.

Basically, Harrod was interested in the possible tension between the conditions required for full employment and those required to achieve a 'steady state of growth'. He employed the Keynesian variables of income, investment, and saving, although he thought of them in growth-rate, rather than absolute, terms. Briefly, his argument was that if growth were to achieve a steady full-employment path, the actual rate of growth (which requires the rate of capital accumulation to equal the *ex post* ratio of saving to income) would have to equal both the rate of growth that would satisfy economically rational entrepreneurs (the 'warranted' rate, or what would now be called a rational expectations equilibrium), and the rate of growth that achieves full employment with no inflation (the 'natural' rate). Any divergence of the 'actual' rate of growth from the 'warranted' rate will, through the acceleration principle, cause a cumulative divergence of an economy's growth path towards an inflationary or deflationary state. Steady, or equilibrium, growth is only achieved if the actual rate of capital accumulation is the rate that keeps entrepreneurs happy and is therefore equal to the actual rate of saving. This is Harrod's famous 'knife-edge'. And further, if the 'warranted' rate of growth is not equal to the 'natural' (or potential) rate – which is determined by the underlying demographic, resource, and technological conditions – the actual equilibrium growth path will not be at the full-employment level. In this scenario the rate of capital accumulation that satisfies entrepreneurs is not equal to the rate that is required to produce full employment with no inflation. Harrod, in Keynesian spirit, thought that the 'warranted' rate would exceed the 'natural' rate, thereby generating permanent unemployment, owing to the inflexibility of the rate of interest produced by Keynes' liquidity trap.[12] He accordingly advocated Keynesian policies to equate the 'warranted' and 'natural' rates.

We need to ask: in what ways did Harrod, and Domar after him, contribute to the development of growth theory? The main positive contribution was to show that simple abstract models could be constructed so as to provide theoretical *outcomes* equivalent to those in the contemporary world, at least when Harrod (but not Domar) wrote his first paper on this topic. It could also be claimed that this was the first growth model to be developed in mathematical form – the first to be truly modern. But the model is extremely inflexible and is only appropriate, in any sense, to real-world situations in which economic change is highly unstable, or those in which continuing inflationary or deflationary tendencies exist. Clearly, it is not a general model of growth, and it tells us little about reality and nothing about historical *processes*. Even an economic technocrat such as H. F. Hahn has said:

> It is also unwise to identify the steady state – say, the steady state rate of growth in output per head – with historical trends in the variable. That would require a good deal more argument than the theories provide. A steady state equilibrium is simply an extension of stationary equilibrium.[13]

It is illuminating that he needed to warn other economists of this obvious danger of equating variables in the Harrod model with quantifiable aspects of reality. Despite the technical precision of this model, and the models that were to follow, it possesses none of the perceptive understanding of real-world processes that is the outstanding characteristic of Schumpeter's model of growth. In this important sense, the Keynesian growth models constituted a step backwards in the history of economic thought. The unsatisfactory treatment of technological change, which totally ignores the achievement of Schumpeter, is another retrograde development. Only in the 'new' growth models of the late 1980s has technological change been treated endogenously, and then only indirectly. Perhaps the best that can be said for these models is that they provided a foundation on which modern growth theory is based, and which some *believe* may ultimately tell us something worthwhile about reality. But that outcome has yet to be demonstrated.

Changes in the real world quickly made the Harrod–Domar model obsolete. The fear of Keynesians that either a deflationary or inflationary gap would emerge in the generation after the Second World War was unfounded. Indeed, during the 1950s and 1960s, Western economies experienced unparalleled rates of growth without significant levels of either inflation or unemployment.[14] While economists thought that this was largely due to the new Keynesian full-employment policies, economic historians, who are more accustomed to the complexities of reality, had strong reservations – reservations that have been amply borne out by the inability of economic policy to prevent the emergence of high rates of *both* inflation and unemployment from the mid-1970s. Despite their confusion regarding the underlying causes of the 'golden age', it was clear to economists that the 'predictions' of the Harrod–Domar model were incorrect. Accordingly, growth theorists such as Swan at the Australian National University, and Solow at MIT, developed so-called neoclassical models in which the equilibrium capital–output ratio adjusts to market forces in order to equate the 'warranted' and 'natural' rates of growth through the interest rate so as to produce a full-employment equilibrium growth path.[15] Swan elegantly demonstrated this neoclassical solution through his well-known growth rate–output/capital ratio diagram.

The Swan–Solow attempt to extend the Harrod model to achieve an equation of the 'warranted' and 'natural' growth paths led, in Schumpeterian fashion, to a vast 'cluster' of journal articles on neoclassical growth models. In the first place, there were attempts to clarify and correct the Swan–Solow solution based upon the mistaken notion that Harrod regarded the actual growth path as an equilibrium path, by N. Kaldor (1955), J. Robinson (1965), L. L. Pasinetti (1962), F. Modigliani and P. A. Samuelson (1966), and J.E. Meade and F. H. Hahn (1966). There were also attempts to extend the simple one-sector model at first to include two sectors – consumption and investment – by H. Uzawa (1961), R. M. Solow (1962), K. Inada (1963), and A. Takayama (1963); and then to include many sectors by M. Morishima (1964) and F. H. Hahn and R. C. O. Matthews (1964).[16] This massive expenditure of intellectual resources on growth models merely provided a technical account of the conditions under which highly abstract and simplified models, involving a limited number of variables such as output, labour, and capital (in various combinations and forms), generate equilibrium growth. While the degree of technical sophistication of these models was far greater than that of Harrod and Domar, they suffered from the same limitations. These models tell us little about reality and they experience the same inflexibility (or lack of generality) as the Keynesian growth models in that they are only 'relevant' to periods when the real-world economy is on a full-employment equilibrium path. The neoclassical model was unable to predict, and appeared totally impotent in the face of, stagflation, which emerged after 1974. And even though technological change was more clearly specified by growth theorists it was not treated endogenously but rather grafted awkwardly on to the body of these growth models.

Therefore, while both Keynesian and neoclassical growth theories were capable of being adjusted to generate different *outcomes* – of both divergence from and convergence to an equilibrium growth path – they tell us nothing about real *processes*. In fact, even one of the participants in the development of this theory has said:

> Neoclassical growth theory is not a theory of history. In a sense [that is, in its focus upon longrun equilibria] it is not even a theory of growth. Its aim is to supply an element in an eventual understanding of certain important elements in growth and to provide a way of organising one's thoughts on these matters.[17]

He goes on to say that the theory provides 'a good base camp, for sallies into the study of particular economies'. This, of course, is a matter of faith rather than fact. I know of no empirical study of historical growth *processes* that has employed either the Keynesian or neoclassical growth model as an analytical device. It is possible, however, to nominate an outstanding study that is based upon Schumpeter's work.[18]

It is true, however, that growth theory appears to have stimulated an interest in measuring growth and its sources, usually referred to as growth accounting. But this empirical work, true to the theory, is interested not in processes but in outcomes, and it tells us little about the way in which economies change over time. Interestingly, there has been little interaction, despite the claim to the contrary by some 'new' growth theorists,[19] between growth accounting and growth theory. Most theorists have been dismissive of the empirical results on the grounds that they contradict the theory!

The 'new' growth models

The late 1980s saw a revival of interest in growth models, possibly because of the problems Western economies had been experiencing over the previous decade in achieving desired rates of growth, together with the theoretical advances that had been made in the fields of technological change and dynamic general equilibrium models. This renewed interest in growth models can be traced back to a paper in 1986 by Paul Romer,[20] which attempted to demonstrate that investment undertaken in an economic system experiencing increasing returns to scale will not necessarily reduce the marginal product of capital to the discount rate. Accordingly, the incentive to invest could exist 'indefinitely', enabling sustained growth in per capita income over the 'longrun'. Apparently, without Romer realising it, this approach cuts across the central stagnation thesis, based upon the idea of diminishing returns in growth theory since the time of the classical economists. But rather than follow the idea of diminishing returns back to the classical model, Romer superficially traces the idea of increasing returns back to Adam Smith's pin factory. In fact, his starting point was not the important issues in the history of economic thought, but rather papers by K. J. Arrow (1962) and Uzawa (1965) on learning by doing and intangible human capital, together with other papers in the 1970s and 1980s on increasing returns. In other words, the 'new' growth theory is not grounded in earlier work on this subject. Is there little wonder that promising work is not built upon in economics, and that the wheel continues to be reinvented?

If the 'new' theory is little concerned with its theoretical predecessors, what about their concern with reality? Romer claims that there is empirical support for the idea of increasing returns to scale leading to growth rates that increase over time, and for large countries to grow faster than small countries. To support this contention he presents three sets of data. First, he claims that the empirical results of growth accounting provide a rough indication that the increase in output is the result of something more than the increase in inputs, and that this in turn suggests that increasing returns to scale exist. Other 'new' growth theorists, however, have rejected the results of growth accounting because they *believe* that they underestimate the true role of technological change.[21] They may be correct (as suggested later in the chapter) but they do not say why their intuition is better than the empirical results. Secondly, he compares the growth rates of developed and undeveloped countries and concludes that while there is convergence within country groups, there is divergence between these country groups, which is further evidence of increasing returns.

Thirdly, he briefly looks at growth-rate data for the USA over the period 1800 to 1980[22] and concludes that growth rates have increased over time, thereby providing additional evidence of increasing returns. Actually, what these data appear to show is that growth rates fluctuated widely even after being subjected to smoothing, and the only reason that the twentieth-century growth rate appears higher than that in the nineteenth century is the excessive weight given to the impact of the Second World War in the 1940s and the golden age of the 1950s and 1960s – these three decades are historically abnormal and may never occur again in such a combination. These data are not good evidence of increasing growth rates, as the long period of slow growth and stagnation since the mid-1970s suggests. Romer also mistakenly assumes that growth is a modern invention: 'A general pattern of historically unprecedented growth for the world economy is evident, starting in the last part of the 1800s and continuing to the present.'[23] Evidence of modest increasing returns – with the sum of the partial elasticities amounting to 1.15 – for the late eleventh century is presented in Chapter 6, and it is shown in Chapter 7 that economic growth is not a modern phenomenon, as rapid rates of growth of real per capita income have been experienced in Europe during the past millennium. As can be seen from Figure 3.2, growth rates were as high as 0.6 per cent per annum in the eleventh century and 1.6 per cent per annum in the sixteenth century. In other words, growth has been a normal feature of human society for at least the last 1000 years, but it has not been continuous, nor has it occurred at an increasing rate as the economy has grown in size. Rather, it has proceeded by long upswings followed by long downswings – great waves of economic change – with each lasting for up to two centuries – see Figure 7.3 on page 257. And within these great waves of change, returns to scale, and hence growth scales, have fluctuated markedly. We just happen to be in a long, but fluctuating, upswing that began with the Industrial Revolution and could well end with the growing pressure of population and pollution on world resources. It is possible that we are on the crest of the modern great wave of economic change.

Hence growth in the real world is far more complex than Romer recognises. Like many economists before him, he is guilty of a form of casual empiricism that is employed merely to rationalise theoretical preconceptions. No attempt is made to understand the historical forces that are transforming human society. Instead they are merely dealing

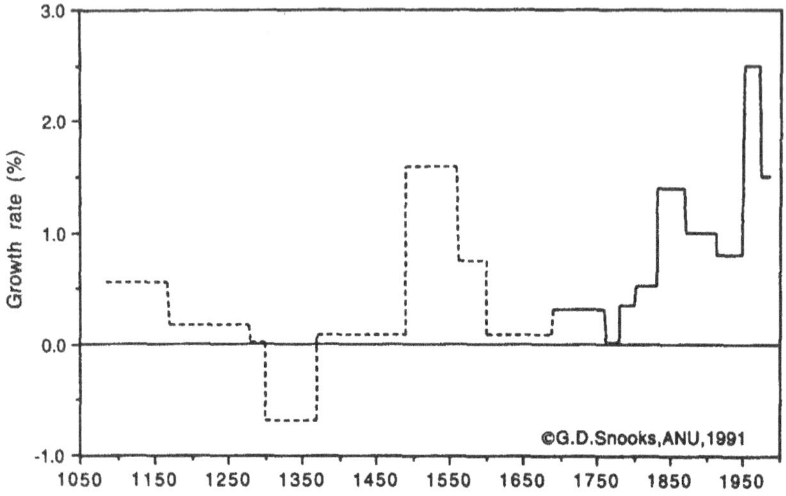

Figure 3.2 Growth rates of the English economy, 1086–1989

Sources: 1086–1688: estimated from simulated variables in Figure 7.3 (see page 257); see Chapter 7, Note 58.
1688–1989: Table 7.3 (see page 247).

with the ripples that pass briefly across the surface of these great economic changes. And in view of the cavalier attitude to data and empirical techniques, one must remain sceptical of the claim by R. J. Barro and Romer in the *NBER Reporter* that 'Two years ago, we created a growth project to encourage the kind of interaction between theory and evidence that once [in the 1960s] had been so successful'.[24] We have already dealt with the myth that there was any 'interaction' between theory and growth arithmetic except for rationalisation purposes, and we have doubts that the reference by Barro and Romer to longrun and cross-sectional growth rate data – which only provide evidence on *outcomes* – can be usefully employed in their stated objectives to examine *processes*. Only historical studies of growth can inform us empirically about *processes*. If data about *outcomes* are being used, the implication must be that their models are also concerned with *outcomes* rather than processes. In Part II a more rigorous attempt is made to consider this issue.

Whatever the inspiration of the 'new' growth theory, it focuses not upon physical capital but on 'knowledge capital' because 'the public good aspects of knowledge as information naturally create increasing returns to scale in many contexts'.[25] In doing so they are attempting to model – even if indirectly through 'knowledge capital' – technological change endogenously. This objective has led them to an investigation of investment in human capital and in the research and development of new technologies. This involves, in their words, 'a rigorous accounting of the resources used up in creating new knowledge, and . . . explicit consideration of the profit motive that drives private investment in *R&D*'.[26] These are issues with which Schumpeter was concerned at the beginning of the century, although his work was not their inspiration.

The 'new' growth theory of the late 1980s may be on the right track. It is concerned with the motivation – profits – that drives the system (modern growth theory has finally caught up with the classical economists!); with the integration of technological change and trade into growth models (moving on from the classical economists to Schumpeter); and with the analysis of the neglected role of human capital in growth theory. Nevertheless, these models are still highly abstract, overly-simplified, and unrealistic, and they have yet to deal with economic processes. Also, the policy conclusions they draw are hardly novel. For example, G. M. Grossman and E. Helpman claim:

> The advances in growth theory enable us to address rigorously many issues that have long been central to international economics. For example, to what extent and in what ways might international trade serve as an 'engine of growth'? Do international exchanges naturally enhance the growth performance of individual trading countries? And what economic policies are especially conducive to high levels of welfare in a growing, open economy?[27]

There is a large literature on these issues in analytical economic history, of which those economists seem totally oblivious.[28] Typically, they are willing to draw policy conclusions directly from their unrealistic theoretical models, without testing them empirically, or consulting the available historical results.

Of what value is growth theory?

For over two hundred years economists have attempted to model the changes that they have casually observed taking place around them. But what have they achieved? Does growth theory provide any insights to help us understand real growth processes in the past, present, and future? And does it suggest any policy prescriptions, beyond those provided by other means (such as historical investigations), that will assist individual nations and the international economy to maximise human welfare? To provide a positive answer to the first of these questions about real growth processes, it is necessary to test growth theory using historical data. Of the theories surveyed in this chapter the most useful in reconstructing historical growth processes, as will be discussed in Chapter 4, is the least abstract – that by Joseph Schumpeter. Schumpeter was concerned with dynamic processes, and like the classical economists he saw growth as a process that embraced the entire economy and society. In contrast, the Keynesian, neoclassical, and 'new' growth models, provide very limited insights regarding real-world processes, precisely because they are so highly simplified and abstract, and because of the very strong conditions that they impose. Also, as we have seen, they are concerned not with *processes* but with *outcomes*.

This raises a more general issue. Deductive theory may be useful in examining particular economic issues at a point in time, but it has little success in examining the way economies change over time. By attempting to explain a particular phenomenon, such as the price level for a particular commodity, at a single point in time, it is possible to abstract from the complex interactions between economic and non-economic forces and institutions, and economic theory comes into its own. Economic growth, however, is a very complex interactive process involving the entire society with all its institutions and value systems undergoing changes that are non-marginal. Such a multi-dimensional process cannot be satisfactorily encompassed by the *ceteris paribus* world of deductive theory. Schumpeter is one of the few growth economists to formally recognise the complexity of real processes of change. In the opening sentences of *Economic Development* he says: 'The social process is really one indivisible whole. Out of its great stream the classifying hand of the investigator artificially extracts economic facts.' Not surprisingly, his theory possesses a richness that is missing in the more abstract work of economists that have come after him. Modern deductive economists assume away both the richness and the complexity that is required to explain the process in which they claim to be interested. In the event they have cast away pearls that are worth more than their tribe.

Clearly, the strength of economics is not in examining change, rather it lies in analysing specific economic phenomena at a point in time and possibly comparing the results with a similar examination at another point in time. These economic phenomena include the operation of markets; the distribution of resources, goods, and income; and the working of the system of production. For example, the modern theory of production can be effectively used (as in Chapter 6) to examine the microeconomic operation of the manorial system in England in 1086, the year of the great Domesday Survey; but it is not a very effective tool when used as the basis of growth accounting over longish periods of time. The reason is that the *ceteris paribus* nature of production function theory holds constant some of the important determinants of growth, such as institutional development involving changes in markets and in the resulting transaction costs. It is doubtful that formal deductive models will ever be able to examine satisfactorily change in human society, particularly over the very longrun.

On this issue, it is not clear what the 'new' growth theory regards as the 'longrun'. It would appear to have very little to do with real time, as is made clear by the wildly ahistorical claim of two of its advocates that, under conditions of increasing returns to scale, 'the incentive to accumulate capital may persist *indefinitely*, and long-run growth in per capita income can be *sustained*.'[29] There is no historical example of investment and growth occurring 'indefinitely'. They seem to be misled by the mistaken view that growth began with the Industrial Revolution and has proceeded more-or-less continuously ever since. As will be demonstrated in Chapter 7, if we take one country with a history unbroken by invasion for a thousand years – England from 1066 to 1991 – we will observe that growth can be considered a normal condition of society, but that it has proceeded through a number of great waves of change rather than in a sustained fashion. Growth has been persistent throughout the very longrun but it has never been 'indefinitely sustained'. It is these great waves, together with variations in returns to scale and growth rates for shorter periods, that we need to explain.

Let me anticipate the criticism that modern growth theory is not concerned with the very longrun. There are two responses that can be made. First, that the very longrun *is* an appropriate test for growth theory, because of the great acceleration that has occurred throughout this century in the rate of institutional and systems change, an acceleration that is likely to continue into the future. It took, for example, 800 years for the English feudal system to be transformed into industrial capitalism. But the institutional and technical changes that have occurred in the last 200 years have probably been greater. It is also sobering to reflect upon the rapidity of the current changes taking place in the entire economic system of Eastern Europe. As greater economic experience is being increasingly compressed into given units of time, the 800 years before the Industrial Revolution can be regarded as a slow-motion version of the type of institutional change that we are currently experiencing and will experience in an accelerating fashion in the future. Not only is the very longrun relevant, but it is more 'readable' because of its slower pace. Secondly, the realisation that growth in human history has proceeded by great waves of change, may be vital to our future. If growth is not 'indefinitely' sustainable, what are the conditions under which the modern upswing is likely to end? This is hardly an irrelevant consideration.

The second question raised at the beginning of this section was whether growth theory can provide policy prescriptions that are more useful than those that emerge from a careful observation, and analysis, of the historical growth process. Classical theory, for example, warned against government intervention in order to protect vested and non-progressive interests (particularly landlords), and championed free trade; Malthus also warned of the problems of population increase and of welfare policies that would encourage reproduction; Marx told of increasing unemployment and immiseration of the working class as capitalists attempted to maintain profit rates, which would ultimately cause the collapse of capitalism; Schumpeter re-emphasised the classical lesson about the importance of competition which, in this case, was required to provide an appropriate environment for the innovating entrepreneur, and he told us about the importance of the process of depression in preparing the way for a new phase of expansion; Harrod emphasised the need for Keynesian policies to avoid a growing deflation; the neoclassicists emphasised the importance of market forces in achieving a full-employment equilibrium growth path; and the 'new' growth theorists have so far emphasised the importance of trade, innovation, and selective government intervention to maximise the welfare of countries producing export goods and investing in research and development. Of these policy prescriptions, the most original and helpful are those suggested by Schumpeter and the classical economists. The remaining suggestions do not appear to go beyond anything that either static economic theory (in reference to Harrod) or analytical economic history (in reference to the 'new' growth theorists) has told us.

APPLIED ECONOMICS

Growth accounting

Growth theory, therefore, cannot be usefully applied to reality because it is highly simplified and unrealistic and it does not deal with economic processes. These models, as models of growth, have not been, and cannot be, set up in a testable way. The best that can be done is to test the predicted real world *outcomes* of these models – such as diminishing returns and stagnation in the classical models; deflation or inflation in the Harrod–Domar model; convergence of growth rates for countries with similar preferences and technologies in the neoclassical models; or increasing returns, increasing growth rates over time, and divergence between growth rates of groups of countries at different stages of development. Possibly the only exception, once again, is Schumpeter's model, which deals with processes as well as outcomes; but even this model can be applied only in a roundabout way; it cannot be tested directly as it has not been set up in a testable manner, and if it had it would probably lose much of its explanatory power. Processes of economic change, as yet, cannot be captured in simple deductive models.

The closest that the profession has come to applying economic theory to the issue of growth is in the field of growth accounting. Not surprisingly, growth accounting suffers from the same deficiencies as growth theory – it is curiously static, dealing not with processes but with outcomes. The theoretical foundation of growth accounting can be traced through the growth model of Solow (1957) to the work of P.H. Douglas and C.W. Cobb (1928) and to P. Wicksteed (1894),[30] and employs a production function of the form:

$$Q = AL^{\alpha}K^{\beta} \tag{1}$$

where Q is the maximum value of output that can be produced with the quantities of labour L and capital K; α and β are the partial elasticities of labour and capital respectively, which sum to one; and A is the measure of efficiency of production. From this production function the following total factor productivity index has been derived:[31]

$$\dot{A} = \dot{Q} - (a\dot{L} + b\dot{K}) \tag{2}$$

where $\dot{Q}$, $\dot{L}$, $\dot{K}$, represent the rates of growth of output, labour, and capital respectively, and $\dot{A}$ is the rate of growth of the residual which is interpreted as including technological change together with other changes in productive efficiency; and a and b are the factor shares in value added of labour and capital respectively. This index is employed in growth accounting to estimate the sources of economic growth.

Conceptual problems

It is well known that this type of total factor productivity index is plagued by a number of important problems. For example, the underlying production function makes a number of very strong and unrealistic assumptions. First, it assumes that factors are paid their marginal products, which is only true under perfect competition, a situation that is rarely met in reality. Secondly, the function assumes constant returns to scale, which satisfies the predictions of neither the growth theorists from Adam Smith to the neoclassicals, who assumed decreasing returns, nor the 'new' growth theorists, who assume increasing returns. Thirdly, the Cobb–Douglas function makes the very restrictive assumption that the elasticity of substitution between resources is equal to one (that is, that iso-quants are log-linear in shape). Although more flexible forms (CES) have been developed they are less popular than Cobb–Douglas because of their greater estimation problems.[32] Fourthly, a less well-known problem is that the geometric index, as specified above in (2), behaves erratically and inconsistently when the data fluctuate widely (that is, when the coefficient of variation exceeds 50.0), such as an economy in crisis (Western Europe during the Black Death of 1347–49, or Eastern Europe today), or at the industry and firm levels.[33] In such cases it is more appropriate to employ the following arithmetic index:[34]

$$A = Q/aL + bK \qquad (3)$$

While there is a theoretical foundation, albeit somewhat shaky, to the calculation of total factor productivity, much of the subsequent work from 1962 by growth accountants is largely pragmatic and *ad hoc*. E. F. Denison is the main exponent of growth accounting and he has taken the technique to its logical conclusion by including (using factor shares as weights) every conceivable quantifiable variable that could possibly contribute to a change in real GDP. He does not appear at all concerned with the theoretical implications of this procedure. Recently (1987), he maintained:

> Growth accounting starts by recognising that many different determinants govern the size of a country's output at any given time. It deals in the first instance with direct determinants of output such as the number, hours, demographic composition, and education of employed persons; quantities of land and capital; the stock of knowledge; the size of markets; the extent to which actual practice departs from lowest cost practice; the amount by which resource allocation departs from output-maximising allocation; and the intensity with which factor inputs are used.[35]

The theoretical justification for combining the main factors of production are clear, even if the assumptions on which it is based are very restrictive, but it is less clear that there is any theoretical justification for the extension of this technique to include a large ragbag of adjustments, such as safety legislation and various questionable schooling variables. Denison attempts to justify this *ad hoc* procedure by invoking 'the principle of proportionality', which has no economic justification. Few theoretical economists seem happy with this pragmatic methodology.[36] Indeed, this approach is dangerously close to being 'measurement without theory'.

Practical problems

In addition to the theoretical qualifications, there are a host of criticisms that can, and should, be made concerning the quality of the statistical estimates of both output and the various inputs. First, there are questions that need to be resolved about the appropriate measure of output. The major issue here concerns the inconsistent nature of GDP over time because of the changing balance in an economy between the market and household sectors.[37] This is aggravated by problems concerning the changing relative size of both the underground economy and rural subsistence production. Secondly, we need to contend with the problem of defining and measuring the stock of capital services.[38] Thirdly, raw labour-supply figures are only an approximation of labour services. Finally, and possibly most important of all, price deflation techniques are essential but heroic, particularly concerning choices of weights, base years, and problems of quality change. While these issues arise in all applied work they are particularly critical in the calculation of the 'efficiency of production' because of the residual method employed, as all measurement errors are captured in the residual. Presumably one could keep tinkering with an infinite number of adjustments until the residual disappeared. It is possible that the detailed adjustments undertaken by Denison impose a greater burden than can be borne by the validity of the theoretical underpinning, the quality of the data, and the precariousness of the statistical manipulations. A better course of action might be to apply a simple index of total factor productivity to data on output and the main economic resources (land, labour, and capital), to establish an approximate quantitative framework for the sources of growth as a preliminary step to a more empirical analysis of technical change and economies of scale. Denison claims too much precision for the technique.

Achievements

Growth accounting is important because it establishes a quantitative framework for sorting out the supply determinants of growth. It provides a starting point – by suggesting what we should be looking for and where we might find it – for the analytical and quantitative study of growth and productivity change. But it is only a starting point. Although the work of growth accountants – including Denison, Matthews, Feinstein *et al.*, Ohkawa and Rosovsky[39] – focuses upon the longrun of between 50 and 120 years, they tell us little or nothing about growth *processes*, only about growth *outcomes*. We are told how rapidly the economy has grown and what proportion of that growth rate can be accounted for by changes in capital stock, human capital, labour services, land supply (through reclamation or destruction), economies of scale, and technology; but we are not shown the process by which these outcomes are achieved. Growth accounting is an important, indeed essential, tool, but it is only a beginning for the study of the historical process of economic growth.

Even as a tool for the examination of historical processes, however, it possesses significant problems. The production function theory (Cobb–Douglas) on which growth accounting is based is a static concept which is most meaningful when applied to cross-sectional data. It does not cope well with time-series data because of the severely distorting assumptions that it imposes. The argument involves two issues. First, the growth accounting technique assumes that relative factor prices are unchanged, that there is no substitution of factors over time (that is, that technological change is neutral), and that the growing size of the economy does not involve increasing or decreasing returns to scale. While only marginal changes can be validly handled by this technique, changes over two to five generations are rarely marginal. Far greater sophistication of the theoretical framework is required, by including an interactive term between capital substitution and increased efficiency, and by allowing for variable returns to scale. Secondly, the *ceteris paribus* nature of production theory, particularly concerning the changing institutional structure of the economy under study, further distorts any measurement of productivity. When dealing with periods of time that encompass between two and five (or more) generations, we need to consider the impact of institutional change on productivity. As suggested in Chapter 7, in the very longrun, institutional – even systems – change may be a major determinant of growth through changes in transactions costs.[40] But all this is assumed away by growth accounting, as indeed it is by modern economics. Like deductive theory, growth accounting is a beginning but not an end.

POLICY

Policy instruments: past and present

The greatest difficulty faced by leaders of industry and society is the need to make key non-marginal decisions – decisions involving large numbers of people and large amounts of capital, and possessing the potential for major changes in society and environment. Needless to say, this is not a modern problem – it has existed since the beginning of large-scale tributory civilisations some 5000 years ago – but there is an example *par excellence* in the USSR today (1991). While the rewards for making correct decisions are high (power, widespread acclaim, national progress, and personal wealth), the consequences of making wrong decisions (the loss of power, general vilification, both national and personal bankruptcy and, in extreme cases, death) are intimidating. Rational and successful decision-making at any level is difficult, but on a national scale it can also be highly risky and dangerous.

The central problem of macro decision-making, as is well known, arises from the large number of variables involved, and the difficulty of predicting the impact on those variables of intervention. The larger the scale of the issue, therefore, the more information is required to undertake sensible decisions, the more complex is the process of computing the correct outcome, and the more things there are to go wrong. And these problems increase exponentially with the future time involved. This is the cross that all leaders have had to bear throughout human civilisation. It is instructive, therefore, to review briefly the ways in which societies up to the present have attempted to handle these problems. This can usefully be done by focusing upon the macro decision-making methods over the last few millennia. As far as I am aware this has not been attempted before.

What has been done before is to survey the origins of the modern approach to economic decision-making by tracing ideas and analytical methods back to Graeco-Roman times. The most impressive attempt to do this is Schumpeter's *History of Economic Analysis*. Despite a great deal of research, Schumpeter was unable to find any signs of systematic analytical economic thought in the extant literature of Western civilisation until the late seventeenth century. He encountered only scattered and fragmentary comments in Plato and Aristotle (who did have views on the economic issues of value, money, and the division of labour), through the largely barren Roman writings, to the writings of the medieval scholastic doctors who had little interest in economic issues apart from the moral implications (such as a 'just' price). With a degree of frustration Schumpeter comments in turn: of ancient civilisations, that 'there is no trace of analytic effort';[41] of the Greeks, that 'they merged their pieces of economic reasoning with their general philosophy of state and society and rarely dealt with an economic topic for its own sake';[42] of the Roman civilisation at its height, that 'its able rulers and the galaxy of brilliant generals and administrators around them could have made use of a brain trust.Yet there was nothing of the kind';[43] and of the medieval scholastic doctors, that 'they groped for their generalisations rather than stated them'.[44]

Schumpeter was able, however, to find evidence for practical ideas and instruction on agricultural, commercial, and financial matters beginning in ancient Greece[45] and Rome[46] and re-emerging in medieval Europe on issues such as public administration from the twelfth century; commercial practice in the trading centres of Europe from the fourteenth century; farm management and rural technology from the sixteenth century; double-entry bookkeeping from the fifteenth century; and numerous guides to finance, trade and commerce in the seventeenth and eighteenth centuries.[47] In other words, there is considerable evidence for the existence of practical manuals that are useful for individuals in private and public life in making marginal decisions, but no evidence until the late seventeenth and early eighteenth centuries for evidence of abstract economic thought that could be employed to make the critically important non-marginal decisions.

Clearly, Schumpeter was puzzled by this. How could civilisations in the ancient and medieval world achieve so much in terms of science, mathematics, literature, and philosophy, and so little in terms of analytical economic thought? Clearly, analytical economics was not beyond their intellectual powers. And how could these civilisations build such impressive and successful economic and social systems, together with their associated bureaucracies, without the guidance of deductive analytical thought? Those civilisations had to make the same crucial, non-marginal decisions that we make, but there is no evidence that they were any less successful than we are. Schumpeter recognises the paradox, but he fails to resolve it. He fails, despite his appreciation of history, because he employs the inappropriate and dangerous approach of the modern economist – he imposes the perspective of the present upon the past. In this he stands guilty, along with Marshall, of Cunningham's accusation. Accordingly, Schumpeter asks the wrong question. In looking for the antecedents of the modern approach to decision-making, Schumpeter, and all those who have followed him, have overlooked the ancient and medieval approach to an issue about which they were equally concerned.

Schumpeter's question should have been: how did earlier civilisations attempt to make decisions about the big issues of the day? The answer is that prior to the late seventeenth century, civilised society possessed an alternative method to analytical economic thought for making non-marginal decisions, that was sufficiently effective to discourage them from seeking the 'modern' alternative of which they were undoubtedly intellectually capable. While the civilisations of ancient Greece and Rome, and those of feudal and medieval Europe, developed impressive professional bureaucracies comparable to our own, to handle the day-to-day business of government – involving systems to raise taxes, keep account of receipts, spend funds and so on – they employed an inductive approach based upon historical and contemporary data to resolve the larger issues. This inductive and analytical approach to understanding economic reality could be regarded as the progenitor of both modern economic history and economics. In other words, economics is not the parent of economic history, but rather its much younger sibling.

The social context of this inductive approach requires careful interpretation because, at first sight, it appears arbitrary. By failing to look beyond the ritualistic form of this inductive approach economists have missed the simple truth: that it embodied a rational approach to decision-making. Broadly speaking, the instruments of 'social' policy-making in Britain over the last two millennia have undergone at least four main influences, including the Graeco-Roman, the Germanic, the Christian, and finally the analytical economic. Throughout this period the bureaucratic framework has waxed after the invasion by Rome in AD 43, waned when the legions left in the early fifth century, and waxed again following the late Anglo-Saxon kingdom and the Norman Conquest.

The Graeco-Roman influence

In ancient Greece and Rome, political leaders and leading citizens who needed to make important decisions concerning the state or economy, would approach an oracle for advice and, in very rare circumstances, for prophecy.[48] This approach to decision-making was brought to Britain by the invading Romans in AD 43 and was employed until 313, when Christianity, one of the Eastern religions that were growing in influence in Rome, was adopted as the official religion of the Empire – or more probably in 391, when an edict by Theodosius closed all oracles and forbade divination of any type. It was at this time that the famous Delphic Oracle, which had flourished for a thousand years, was finally abolished.

It is important to realise that the age-old procedure of consulting oracles when making important decisions concerning military, political, and economic matters, was not as arbitrary as it may sound – indeed generally not arbitrary at all. Judgements of irrationality, which are incorrect, arise from a lack of knowledge of the procedure, an unsound perspective based upon the modern use of deductive model-building, and a failure to relate successful outcomes to the procedures that produced them. There were two dimensions to this procedure: the skill and knowledge of the oracle; and the way in which this advice was used by the client. The oracles had to be highly skilled, not only in the mysteries of their craft but also in knowledge of their society, both past and present. When interpreting the signs (divination was either induced from nature or 'inspired' by the gods), therefore, the oracles brought their expert knowledge of society into play. As the income, as well as the fame, influence, and possibly the life of the oracle and attendant priests depended upon the success of the advice given, it could not afford to be either arbitrary, and therefore unsuccessful, or too ambiguous, and therefore meaningless. But a degree of ambiguity was probably not only necessary (and here it was equivalent to the *ceteris paribus* character of modern prediction) but also desirable. While the advice of the oracle contained information relevant to the problem in question, a degree of ambiguity would enable the client to use it in a creative way. This procedure was useful in isolating, and focusing the mind upon, the main issues involved. It appears to have been an intellectual device to assist clients (who included both private individuals and leaders of a city or state) to form opinions about difficult problems and to provide them with the confidence, which was derived from the status of the oracle and her god, required to translate the decision into action. Confidence was (and still is) probably as important as the information provided in this approach. This does not, of course, rule out those more cynical exercises in which a leader, who had previously decided upon a particular course of action, publicly (and possibly corruptly) consulted the oracle in order to convince his followers. The point I wish to emphasise is that the advice of the oracle was not arbitrary in intent or effect, and that it was not blindly followed. It was part of an inductive procedure in which historical and current data were employed in a process of creative thinking to arrive at a sensible decision. I will argue that the obvious success of this approach resulted from the imaginative way in which primitive instruments of policy formation were employed. There appears to be an inverse relationship over time between the quality of the instruments and the quality of the decision-making process – as the instruments have improved, the quality of the decision-making process, including the degree of real-world information, has declined.

This general discussion can be given historical content by examining how the most famous and enduring oracle of the ancient world – the Delphic Oracle, dedicated to the God Apollo – operated.[49] The fact that the Delphic Oracle dispensed advice for at least a thousand years, until abolished by Theodosius in 391, is a measure of its usefulness and success. City elders and individuals would, on the payment of a fee (*pelanos*), seek advice from Apollo through a female medium (*Pythia*) assisted by male priests.[50] Public enquiries from a city appear to have attracted a fee some eleven times greater than that from an individual, and grateful clients were known to shower gifts upon the oracle. It is thought that the priests at Delphi maintained an archive (*zygastron*) which included copies of past advice, and that they gained information about the Greek world from their library and from the many visitors who came from far and wide. Indeed, the priests, whose motto was 'know thyself', were known for their learning and wisdom, and their advice was recommended by both Socrates and Plato. Also, Plutarch was not only a philosopher of importance but also the Chief Priest at Delphi from AD 85 to 125. As one commentator has said: 'When Euripides says "the good prophet is a man skilled in conjecture", he is implying that inductive divination depends less upon revelations from on high than upon the resources of an astute and subtle mind.'[51]

Once the fee had been paid and the preliminaries undertaken, the Pythia and the priests interpreted Apollo's advice and wrote down the response, sometimes in verse, for the client. All serious decisions in ancient Greece, and Rome, were made in this way. The success of this institution is reflected in its longevity, high status, and wealth. No arbitrary decision-making process could have achieved the widespread and long-held confidence of the ancient Greeks and Romans; clearly, the priests employed their knowledge of the contemporary world and of the history of the ancient world to provide effective advice. In fact, history was a major intellectual activity in ancient Greece, and in Rome it was almost an obsession.[52] We know that unsuccessful oracles became the subject of public scorn, lost influence, and fairly quickly ceased to exist.[53] Of course, mistakes were made by the oracles, and clients could, and did, dismiss advice that did not suit their purpose.[54] But, generally speaking, this inductive system seems to have worked just as effectively as modern decision-making procedures based upon deductive techniques. Indeed, it has been widely perceived to be so effective that those preparing for entry to the English public service, until very recently, were trained in classics rather than economics; a tribute to the success achieved by the ancients in shaping their world.

The Germanic influence

At the time of the Roman conquest of Britain, Germanic tribes in Europe employed a functionally similar, although structurally different, technique for social decision-making. Pagan priests employed for this purpose the ancient German alphabet (or 'futhark')[55] which, unlike our modern alphabet, had symbolic meaning. The divination procedure was undertaken using runes carved on fragments of wood from the sacred yew tree for the casting of lots which were then subjected to interpretation by the priest or rune-caster. This method of providing advice was observed and recorded by reliable Roman witnesses, including Plutarch in *Marius*, Julius Caesar in *De Bello Gallico*, and, particularly, Tacitus in *Germania*. This Germanic practice was introduced into England from the fifth century by the Anglo-Saxons.[56] Once again this method should be seen as an aid to rational decision-making which derived authority from the northern gods, rather than as a substitute for it. The pagan priest was an expert not only in the interpretation of runic symbols, but also in the oral history, myth, and legend of Anglo-Saxon society. And he acquired considerable knowledge of the contemporary world from those who consulted him. Only the priest who got it right consistently was able to prosper, and consistent success could only come from the accumulation of experience and a rational approach to available information about the real world. Religious practices merely gave clients confidence in the advice offered – the confidence required to make difficult and risky decisions. In a similar way, modern economists derive authority from the 'mysteries' (as far as the general public are concerned) of their craft – of theory and statistical technique expressed in mathematical symbols that only the initiated can interpret.

Medieval Christianity and bureaucracy

With the return of Christianity to England in AD 596, pagan methods of decision-making and advice were gradually replaced by Christian methods.[57] Christian leaders sought guidance through prayer, by visiting holy relics and holy places, but mainly from their chaplains. Like the oracles and pagan priests, Christian priests were well versed in the knowledge of their times: not only in theology, but also in history, mathematics, and science, so that they were able to interpret God's 'will' in an informed and relevant way.[58] To have relied upon irrational inspiration rather than informed critical thought would have led to disaster and fairly rapid dismissal. The priests' reputations, wealth, and possibly even lives, depended upon getting it right most of the time.

As the role of government became more specialised in the later Anglo-Saxon period, the king, and perhaps some of his more powerful earls, employed experts in their households who devoted themselves to raising funds and organising state, church, and army. An excellent example of this growing specialisation of the king's household and the emergence of bureaucratic functions is provided by the English Exchequer in the Anglo-Norman period.[59] Prior to the Norman Conquest in 1066, the king's funds, or treasure, were supervised by a number of members of his chamber known as chamberlains (or servants of the wardrobe). Initially, they were probably laymen, but with the emergence of a comprehensive system of accounts for the receipt and disposal of the treasure, which was held in Winchester Castle, at least one educated person, usually one of the king's chaplains, was required in this role. He became known as the treasurer, and had a detailed knowledge of the sources and uses of the king's income. These details were recorded by cuts and notches on wooden tallies, an ancient method of accounting that continued until 1834.[60] In fact, the early name for the Treasury – which also held documents used for revenue purposes including the hidage lists, Domesday Book, and the pipe rolls – was the Tallies.

By the end of the Anglo-Saxon period, therefore, Christian priests in the households of the king and great earls combined the age-old role of informed mystic with the more recent technical skills of the financial controller and adviser (which had also existed in the ancient world). This evolving bureaucratic system underwent, in quick succession, two revolutionary changes, one affecting the data base, and the other transforming the system of numerical computation. These changes were as important in the late eleventh and early twelfth centuries as the modern census and the electronic computer are today. The first of these was the Conqueror's great economic survey known as Domesday Book – to be discussed further in Chapter 5 – which provided an incomparable data base for the Treasury for centuries to come. The second was the introduction in the reign of Henry I (the Conqueror's youngest son) of a new institution called the Exchequer, which revolved around a new system of calculation (the Exchequer table or abacus), a system of receipts for revenue received (the tally sticks), and a written record to be housed in the Treasury (the pipe rolls).[61] While the individual elements in this system may have been used before Henry I's reign, it was their combination into a major new method of auditing the financial accounts of the English state that constituted a major step forward in bureaucratic expertise. This new system is thought to have been based upon Arab learning in mathematics and arithmetical calculation, filtered through the famous French school at Laon under the supervision of Master Anselm, famous for his biblical teaching, and his brother Ralph, who was skilled in arithmetic. Many English clerics, including Adelard of Bath, who became an expert on the abacus, attended this school in the early twelfth century.

This reorganisation of the king's finances resulted in the separation of the Exchequer, which handled the accounting system and the associated judicial work, from the Treasury, which henceforth limited itself to the storage and payment of money and other valuables. The name Exchequer was derived from a chequered table cloth (about 10 ft by 5 ft) that was employed as a form of abacus, with counters being placed in the squares by the official 'calculator' to calculate the revenues due from each county sheriff together with monies paid by him. Anyone who has attempted arithmetical exercises, particularly multiplication and division, in Roman numerals (which do not allow for zeros) will appreciate the importance of such a system where absolute accuracy is required, and seen to be achieved by the various representatives of the king. As with the compilation of Domesday Book, there was no opportunity to falsify the record and defraud their royal master. But this was more than just a record of the king's finances, it was an instrument of government. Richard Fitz Nigel, the king's treasurer in the late twelfth century and author (in 1179) of the famous handbook on the Exchequer known as the *Dialogus de Scaccario* (Dialogue of the Exchequer), makes this quite clear when he explains wryly:

> We are of course aware that kingdoms are governed and laws maintained primarily by prudence, fortitude, temperance, justice and the other virtues, for which reason the rulers of the world must practice them with all their might. *But there are occasions on which sound and wise policies take effect rather quicker through the agency of money.*[62]

The English treasurer in the late twelfth century (who was also the Bishop of London) makes it quite clear what he thinks the motivating force in society is: money is more effective than virtue as an instrument of government. There also can be little doubt that monetary policy was being employed consciously, and apparently effectively, in feudal society, despite a recent statement by policy-oriented economists that monetary policy dates back only to the early eighteenth century.[63] Monetary policy is probably as old as the institution of the royal Treasury, which can be traced back at least to ancient Assyrian and Egyptian kingdoms,[64] some three thousand years ago. Needless to say, these instruments of government continued to develop throughout these three millennia in terms of technical sophistication, as has the economy itself. But whether they have become more effective is a moot point.

During the medieval and early modern period, the increasing sophistication of fiscal instruments of government, and the growing secularisation of public policy, were responses to the increasing complexity of both state budgets, particularly during times of war, and the structure of economy and society. As the complexity of the task grew, the difficulty of decision-making increased, generating a need for greater specialised 'economic' knowledge. In addition knowledge about, and attitudes to, the world and the way it works were changing. This helped to undermine confidence in the age-old inductive methods, and prepared the way for a new deductive approach to problem-solving and policy formation. In part this need for specialised knowledge was met by practical manuals such as the *Scaccario* and the many others that followed, and in part by crude attempts to generalise about economy and society. The triggers for these attempts were critical events such as large-scale war, which increased the magnitude of fiscal problems. We can see this in the Conqueror's reign when the need for war finance led to the compilation of Domesday Book, an early form of national accounting. To produce such a document, theoretical knowledge about the nature of the manorial economy, and about the relationship between taxation assessments and economic capacity to pay was required. The next major step was made in the late seventeenth century by Petty, Davenant, and King. They took national accounting a step further (after a lag of 600 years!) by drawing up tables of national income and by generalising about taxation and income, once again in response to war (which also stimulated the officially sponsored national accounting during the 1940s). This practical and inductive response to problems of state finances – the traditional methodological approach stretching back through the millennia – appears to have been the starting point for the emergence, from the early eighteenth century, of a more general and formal body of economic ideas. Over the following three centuries these inductive generalisations were transformed into a body of abstract, deductive thought, which today provides the main foundation for economic policy – a foundation largely divorced from the historical knowledge of earlier government advisers.

The modern period

Modern economic policy prescriptions are typically based upon deductive economic theory. Economists claim that theory provides insights concerning the relationships underlying reality even though it is not descriptive of reality, and that these relationships (largely static) tell us what to expect in the near future. Accordingly, policy implications are drawn either directly from theory, or indirectly through a process of empirical testing and empirical estimation of the parameters of models employed for predictive purposes. But the testing procedure merely confirms that one model is more appropriate than alternative models, not that the preferred model fully captures the real causal process involved. This can only be achieved by a longrun analysis of the economy – an analysis that goes beyond merely tracking the rise and fall of theoretically selected variables over the last generation, which is an approach followed by economists' 'historians'. Longrun analysis is not a substitute for deductive model-building but it is an essential complement to it.

Economists involved in policy-making generally fall into one of two broad categories: those who ignore the past and base their policy recommendations solely upon their favoured models, however selected; and those who test the validity of their models and estimate their parameters by employing data from the near past. The theoretical journals are a good example of the first type: no matter how esoteric and unrealistic a theoretical paper may be, the author apparently feels under an obligation to outline the policy implications. This is a curious ritual derived from the distant past when economic generalisation was developed specifically to resolve a critical economic problem. One hopes that no one except their author takes these recommendations seriously. The other group contains a variety of approaches, ranging from the use of sophisticated econometric methods to estimate the parameters of simple models, using a handful of years, to those who merely track selected theoretical variables over a limited time period in order to arrive at (mis)informed guesses about future outcomes. The problem with this approach is that the variables measured are those suggested by simple static models that neglect interactive processes and are likely to be invalidated by employing atypical benchmarks such as the 1950s and 1960s. But whatever the method, the perspective will be limited and possibly skewed, because the models are merely simple constructs of the human mind.

In concluding this discussion of the instruments of policy, it needs to be emphasised that although the technical sophistication of bureaucracy and economic analysis is far greater today than in the more distant past, knowledge by policy advisers about the nature of economy and society is much less. This is of considerable significance because the quality of decision-making is determined more by the intelligence, imagination, and knowledge of the policy adviser, than by formal decision-making instruments. If this was not so, how did previous ages, in which large systematic errors invited extinction, cope as well as they did? There is no evidence, in terms of outcomes, that they were less able than modern societies in planning and executing large-scale political, social, and economic operations. Policy advisers in earlier societies were able to overcome their limited, and apparently unrealistic, technical apparatus, by their more general skills and knowledge. Today, policy advisers rely too heavily upon their improved technical apparatus, not always realising that its claim on reality is tenuous. Theory is normally used as a substitute for the information about reality that is needed for sound decision-making, rather than as a way of organising that information. By overrating the usefulness of their theory, deductive economists have felt justified in ignoring history.

Policies for growth and stability

To characterise how mainstream economists go about the task of formulating policy, I will continue the growth theme of this chapter. This section demonstrates how historical data are either completely ignored or incorrectly employed.[65] Policies for growth and stability can be treated separately, although there are times when governments employ counter-cyclical policy as if they were attempting to influence growth outcomes.

Policies for growth

The attitude to growth and how best to achieve it has changed dramatically during the post-Second World War years. Immediately after the war, allied governments were concerned to rebuild the European economy as quickly as possible, consistent with the achievement of internal and external stability. The fear, expressed in various international forums such as the Food and Agriculture Organization (FAO) and the International Monetary Fund (IMF), was that the unemployment and stagnation of the inter-war years might re-emerge, or alternatively that overfull employment and inflation might appear. This concern, together with the confidence achieved by economic bureaucrats both practically (owing to experience with wartime controls) and theoretically (because of the persuasiveness of Keynesian economics), encouraged the desire for greater intervention in capitalist economies in the form of 'indicative' planning. While none of the Western democracies wished to go down the Soviet road of a centrally-determined system, they were impressed by the ability of the USSR to produce a formidable war machine capable of crushing Hitler's armies in the east.

Indicative planning's rationale was the achievement of an equilibrium growth path at full employment with no inflation. To achieve this outcome the planning authority would gather information from economic agents and establish a set of feasible growth targets for the economy and its main sectors, and would estimate the investment, labour, and materials required to meet those targets. It was thought that this type of planning would increase the supply of relevant information to economic agents, and reduce the risk of decision-making, thereby increasing the efficiency of the growth process of capitalism. The theoretical basis, or rather *ex post* rationalisation, was the Arrow–Debreu general equilibrium model of welfare as developed by J. E. Meade.[66] But the assumptions were restrictive: that there are no shocks to the system; that no incorrect information is provided by economic agents; and that the reduction of risk was desirable (hence removing the Schumpeterian well-spring of progress?). Also, simple static equilibrium theory is not a good guide to dynamic circumstances – change, as Schumpeter was at pains to point out, involves a process of disequilibrium.

It is interesting that Harrod also supported indicative planning, because he thought it would help to resolve his knife-edge problem. If technocrats could estimate the full-employment (or natural) rate of growth, economic agents would be able to adjust their preferred (or warranted) rate of growth to converge towards it. Hence there would be an optimal allocation of resources over time. The underlying view of bureaucratic efficiency and human psychology is particularly naïve as the planners in both Western and Eastern Europe discovered to their cost. It is an excellent example of the adoption of ineffective and costly interventionist policies on the basis of an unrealistic theoretical model.

The most serious attempt at planning in Western Europe in the post-war years was made by France, a country long experienced in government intervention, during the period 1947 to the late 1970s. French planning began just after the war (the Monnet Plan, 1947 to 1953) in order to rebuild key industries and to co-ordinate Marshall Plan aid as effectively as possible. Planning was undertaken by the *Commissariat au Plan* and involved making forecasts of growth rates for the economy and its main sectors, together with the resulting requirements for inputs. The completed plan had to be reviewed by both houses of parliament, which was then published by the government and made available to the public. While the plan had official sanction, it was not legally binding on either the government or the private sector; rather the purpose was to provide information to economic agents. It became known as indicative planning. The sixth plan, ending in 1975, was largely a forecasting exercise, and thereafter even the targets were dropped.[67]

Although Britain had less prior experience with this type of government intervention, the problems involved in reconstructing a war-torn economy, and the electoral success of the Labour Party in 1945, led to the first British attempt at planning, initially under an Official Steering Committee and, in 1947, a Central Planning Board. This attempt did not win electoral approval, and the incoming Conservative government in 1951 abandoned this initial attempt at planning. But the 'stop–go' policies of the Conservative government were not able to ensure steady growth, and the inflation and export stagnation of the early 1960s, together with changes in the party's leadership, led to an unsuccessful Tory attempt at indicative planning between 1962 and 1964 under the National Economic Development Council (appropriately known as 'Neddy' to 'Goon Show' fans). It was, however, just a mask for the continuation of 'stop–go' policies. In 1964, the new Labour administration attempted to revive planning under the Department of Economic Affairs, but its first plan (1966–70) was largely wishful political thinking, as well as being poorly, and inconsistently, constructed. Not surprisingly, it was officially abandoned as early as July 1966. Planning in Britain was not successful.

Since the late 1970s the tide of opinion about how best to achieve economic growth has gone through a complete pendulum swing. With the economic problems of the mid-1970s, the emergence of a 'new right' philosophy in much of the Western world, and the election of political parties that adopted this viewpoint in the UK (Margaret Thatcher), USA (Ronald Reagan), Australia (Bob Hawke), and elsewhere, previous ideas about formal planning were swept away. Instead, democratic 'mixed' economies attempted to reduce the degree of government intervention by deregulating financial and labour markets and transferring ownership of large public utilities to the private sector. The underlying philosophy was similar to that of classical economics, that the most efficient allocation of resources and the fastest rate of national growth could best be achieved by profit-maximising economic agents operating in an unencumbered market system. More surprising, at least to those who did not recognise the intolerable strain placed upon economy and society, was the collapse of the centrally-determined systems in Eastern Europe.

Despite the rhetoric of the 1980s, deregulation was more apparent than real in capitalist economies. While some markets were deregulated (particularly financial), others experienced either little change (labour), or an increase in regulation (import-competing and uncompetitive export industries), both because of the realities of pressure groups in democratic politics and the perceived need to maintain external balance. Also, while attempts to provide a long-term indicative framework were abandoned – although the pretence was maintained longer in France than elsewhere – control of traditional instruments used to influence growth were maintained. These policy instruments included immigration, tariff protection, foreign investment, domestic saving, education and training, and research and development.

The central question is: what role did economic theory play in growth policy? Theory had an influence at a number of levels. First, the tide that swept central planning away was informed, even if only partially and inadequately, by the changes occurring in the economics profession that have been discussed in Chapter 1. At least, these theoretical developments were used to rationalise the political and economic changes that ambitious leaders wanted to implement. These changes included the growing dissatisfaction with orthodox economics. Until the mid-1950s that orthodoxy had largely been Keynesian, but thereafter it became known as the neoclassical synthesis pioneered in the USA by Paul Samuelson, Franco Modigliani, James Tobin, and Robert Solow, and in Britain by John Hicks. Of course, there were centres in Britain, namely Cambridge, where the post-Keynesians went on the attack against the new neoclassical economics but, despite this rearguard action, the neoclassical synthesis came to dominate the economics profession throughout the English-speaking world by the end of the 1960s. The neoclassical synthesis was based upon two basic assumptions: that economic agents were rational; but that markets were not always competitive and took time to clear. Because of the second assumption they advocated Keynesian policies to maintain full employment. While there has always been a tension between the Keynesian inspired macroeconomics and the neoclassical microeconomics in this synthesis, the cracks did not begin to emerge until the oil crisis of the early 1970s and the stagflation from the middle of that decade. This crisis in the profession gave rise first to the 'new classical' macroeconomics, which claimed that rational agents and non-clearing markets are fundamentally incompatible, and proposed to treat markets 'as if' they were competitive and able to clear instantaneously; and secondly to a revamped monetarism, which preached non-intervention and wanted to assert the superiority of monetary over fiscal policy. Both theoretical developments provided a basis for political policies of non-intervention that swept through the Western world during the 1980s.

Secondly, it is important to understand the use made of the vast literature of growth theory that emerged during the 1950s and 1960s. It is a body-blow to the prestige of economics that the policy pay-off on this theoretical work has been negligible. The Harrod–Domar model suggested that steady full-employment growth would be very difficult to achieve, and advocated the effective use of Keynesian policies in a planned framework. This merely reinforced the shortrun recommendations of the *General Theory* and added no new suggestions. The early neoclassical, or Swan–Solow, model suggested that markets would equate the actual and warranted rates of growth and hence achieve steady growth, although Keynesian policies might be necessary to eliminate longrun unemployment. While the rate of return on the considerable investment in growth theory was, therefore, insignificant, it was not surprising because this theory is highly abstract, specialised, unrealistic, and unsuitable for the reconstruction of historical growth processes.

If growth theory has provided little insight regarding both the way economies grow, and how to ensure that growth is maximised, what is the theoretical basis for growth policies in Western economies? During the indicative planning phase in Western Europe (and central planning in Eastern Europe), the most useful economic tool was input–output analysis, first developed by Wassily Leontief at Harvard and Richard Stone at Cambridge from the early 1930s.[68] Unlike macroeconomics in general and growth theory in particular, input–output analysis is sufficiently disaggregated and empirically oriented to be useful in a variety of areas, including growth policy.[69] Like national accounting, this approach to planning was initially developed to answer specific empirical questions, and only later was there an attempt (by Kenneth Arrow, T.C. Koopmans, and Paul Samuelson) to relate it to neoclassical macroeconomic analysis concerning the production function and technological change. While neoclassical macroeconomics possesses an appealing simplicity and elegance, it is the less elegant input–output model that can handle real-world problems concerning the resource and material inputs required to achieve a predetermined rate of growth.

Nevertheless, this empirical approach to development is basically static in approach. Attempts at dynamic modelling have not been particularly successful, and the alternative adopted, of taking a series of snapshots of the economy over time by constructing a sequence of input–output tables for selected years, is hardly a description of dynamic processes. While we end up with a three-dimensional grid, with relationships between goods produced and consumed from period to period being dealt with in a way similar to the treatment of relationships between sectors in a given year, such an arrangement of data – together with the underlying equations – tells us nothing about real dynamic *processes*, only about actual and projected *outcomes*. But unlike orthodox growth theory, input–output analysis does have practical relevance, and can be used as a basis for longer-term planning, both formal and informal, and as a basis, in econometric models, for forecasting.

Despite the massive endeavour of the world's economists, much of contemporary growth policy is *ad hoc* and uncoordinated. Policy concerning immigration, foreign investment, trade, tariff protection, agricultural subsidies, domestic saving, education and training, technical research and so on are handled by separate arms of government and bureaucracy. Not only is there a lack of co-operation between government ministers and their official and unofficial advisers in these separate portfolios, but there is often active competition between them for supremacy as the premier policy-making body. Indeed, it takes an extremely strong head of government to integrate these various separate activities. This, of course, was one of the attractions to economists (although not always to politicians)[70] of indicative planning. Ironically, this political/bureaucratic fragmentation has made it easier for economic theory to play a role indirectly in growth policy. Theory is at its most effective when dealing with smaller, less complicated parts of the economy, using partial and comparative static equilibrium analysis rather than the whole complex economy using general equilibrium or 'dynamic' models. But despite the usefulness of partial analysis of immigration, trade and so on, what is lacking is an overall analysis of the process of change.

A further problem involves the way economists interested in policy prescriptions use time-series data. The problem is two-fold. First, applied economists make a very selective use of time-series data, employing only those variables specified by their simple models. This involves an inadequate coverage of real-world influences. Secondly, the length of time chosen is critical. Economists usually feel uncomfortable delving back more than a decade or so into the past, possibly because they are conscious that the *ceteris paribus* conditions of their models will be violated, but more likely because their knowledge of history is so deficient. An important problem in historical study is precisely where you break into historical processes: and the question that needs to be asked is whether it is a normal or abnormal period? Most often, economists go back to the 1960s and even the 1950s, both because that is when much of the detailed official data on a range of economic variables first became available, and because it encompasses the period of their own personal experience. Whatever the reason, this is a critical problem that affects the work of econometricians who wish to estimate the parameters of their model, and of the more intuitive applied economists who are interested in projecting time trends. The usual choice of a base period – the golden age of the 1950s and 1960s – is seriously misleading. As it is a period more buoyant and stable than any that have been experienced in the last 1000 years, the resulting econometric estimates and analytical conclusions are invariably biased.[71]

Counter-cyclical policies

The immediate public interest, and hence political concern, is usually focused not upon growth, but upon questions of internal and external economic stability – upon unemployment, inflation, and the deficit upon current account. Political horizons in the democracies rarely extend further than three to five years (depending upon the time between elections), and the public appears to discount the longrun heavily. Conflicts at this level between supporters of the longrun and the shortrun are generally resolved in favour of the latter. Needless to say that this preoccupation with counter-cyclical policy is focused only upon the business cycle of three to five years. No attempt has ever been made to take into consideration at what stage we might be in cycles of longer duration which have been tentatively identified by historical economists, such as the Juglar (7 to 11 years), Kuznets (15 to 25 years), and Kondratieff (40 to 60 years) cycles. Even less consideration is given to the great waves of economic change that have been identified in Part II of this study. In other words, the focus of public policy is largely upon the ripples that briefly pass across the surface of the great waves of economic change in human society.

Economists and policy-makers during the 1950s and 1960s congratulated themselves upon their ability to achieve rapid growth and stability – indeed there are some optimists who still like to claim that this golden age was a product of Keynesian-type policies – but they congratulated themselves too quickly. The slow and uneven growth, together with stagflation, after the mid-1970s made it quite clear to all except the economic ideologues, that economics did not have all the answers. In fact, a careful perusal of the business cycle in the 1950s and 1960s demonstrates quite clearly that most of the downturns were largely a result of inappropriate or mistimed counter-cyclical policies.[72] While the focus of counter-cyclical policy during these years was upon fiscal policy, with monetary policy as a back-up, there were those who argued that this was an inflexible system that might be effective during a prolonged recession, but was not sufficiently sensitive for fine-tuning a buoyant economy. Some argued that freely operating markets would do a better job.

The deteriorating economic conditions which emerged after the mid-1970s finally discredited the heavy-handed interventionist policies that had been invoked in the name of Keynes and, as we have seen, the dissenting voices in the profession promoting both monetarism (that advocated free markets but with stable money growth) and the 'new classical' macroeconomics (that emphasised the need for rapidly adjusting markets) began to be heard. In turn, this provided an opportunity for politicians of the 'new right' to present and to pursue a new market-oriented vision of society involving less intervention and a greater role for market forces.

In the main, the balance between Keynesian instruments of taxation and government expenditure on the one hand, and monetary instruments of open-market operations, discount rate, and reserve ratios set by central banks to control the money supply on the other, shifted strongly towards the latter. Accordingly, income tax rates have been reduced (although often replaced by taxes on transactions), attempts have been made to reduce government expenditure, large government businesses (telephone organisations, airways and so on) have been sold to private interests, and there has been a move to deregulate business (particularly banks), and factor markets (with greater success in capital rather than labour markets). Not all these policies have been mutually consistent. In particular, deregulation of capital markets, which has led to destabilising international flows of funds, has made it difficult to achieve stable growth in the money supply. Also, the interventionist impulse is difficult to suppress and government concerns with growing current account deficits have led to the use of high-interest-rate deflationary policies, usually combined with protectionist policies, in an attempt to reduce imports, rather than the establishment of competitive structures that would create greater export potential. Intervention, therefore, has been brought back through monetary policy, and we are faced once again with the 'stop–go' tactics that were rightly criticised during the 1950s and 1960s – evidence of the ahistorical approach of economists and policy-makers.

This *ad hoc* approach to economic management is a bit like making running repairs to a moving machine, without looking where it is going or where it has been. A failure to look back or forward has resulted in a very marked tendency to repeat past mistakes – mistakes that could have been avoided if economics had been an historical social science. An excellent example of this is the commercial, financial, and political excesses of the 1980s that contributed to the recession of the early 1990s in some Western economies. Consider, for purposes of illustration, the Australian case. The recession of the early 1990s in Australia was marked by the collapse of a number of private non-banking financial organisations, the 'bankruptcy' of a number of state government banks (in the Labor-held states of Victoria, South Australia, and Western Australia), the collapse or set-back of the financial empires of some of Australia's largest entrepreneurs (Alan Bond, Laurie Connell, Christopher Skase, Robert Holmes à Court, John Eliott and so on), and the exposure of close political connections with some private speculators. This messy financial, speculative, and political situation critically aggravated the downturn into recession, both directly and through its negative impact upon business confidence. The point that needs to be made is that these complicating factors could have been avoided had economists, policy-makers, financiers, businessmen and politicians been aware of the full cost of their financially unsound, speculative, or ethically questionable activities.

In the case of Australia, there was a very unambiguous example of these costs, namely the speculative, financial, and political excesses of the 1880s, which greatly aggravated the depression of the 1890s and delayed the subsequent recovery.[73] As a result of unsound lending policies by financial institutions in Victoria during the 1880s, large numbers of non-banking financial institutions collapsed, bringing with them numerous speculators and unethical politicians. Finally, in 1893, thirteen of the major banks failed. Although twelve of the banks were successfully reconstructed it was at the expense of public confidence and depositors' funds. The interesting lesson to be drawn is that the younger generation involved in this financial/political/speculative debacle remembered and, as a result, did not make the same mistake during the boom of the 1920s. In fact, the entire financial structure was rebuilt carefully from the ashes and operated upon proper prudential lines;[74] speculators were more circumspect; and politicians, with notable exceptions (such as Edward Theodore in Queensland), were more careful about involvement in questionable business relationships. Accordingly, there was no financial collapse in the Great Depression, and the resulting speculative and political fall-out was quite modest. Collective memory therefore played a positive role in minimising the undesirable effects of the Great Depression. But the thread of collective memory cannot stretch beyond one or two generations without breaking. A hundred years after the debacle of the 1890s, the unnecessary institutional/political excesses have re-emerged. The important lesson here is that we need to institutionalise collective memory by the widespread exposure of economists, policy-makers, financiers, and businessmen to the study of economic history.

CLAPPING REQUIRES TWO HANDS

The timeless approach to economics, therefore, does not come to grips with the real-world, historical processes of change, either in its highly abstract deductive models, or in its attempts to derive effective policy recommendations. While the major predictions of economics are invariably wrong, economists keep employing the same ahistorical models and methods. No attempt is made to change the structure of these models in the light of history. The black-box approach to prediction is just not good enough. Unless, as economists, we understand the underlying dynamics of the economy, we will never be able to predict large, non-marginal changes nor be able to cope very well with them when they burst upon us. Success with marginal, and hence non-problematical and uninteresting, changes should not obscure this fact. The massive amount of human capital that has gone into the development of the economics profession over the last three centuries is wasted if it can only deal with small issues.

I hope it is not too extreme to suggest that economists do not adequately comprehend the forces with which they are dealing. These forces are continuous, proceeding from the past, through the present, and into the future. The world is, and has always been, dynamic and constantly changing. And only by understanding the real processes of a dynamic society can we make sensible predictions and effective policy. Economists are unable to capture this sense of reality and, therefore, without assistance, are not in the best position to help society to achieve its objectives. While clapping – which produces waves of a different kind – requires two hands, the economics profession has one hand tied behind its back. To maximise our understanding of economic reality there must be a close and equal working relationship between theory and history. There is no other way.

4 The Custodians of Real Time

For longest time to him is short

(John Milton, 1671)

COMPETING TRADITIONS

As the economics profession turned its back on real time – particularly in Britain after the 1870s – a new *academic* discipline emerged to carry on the task of reconstructing the past. That discipline was economic history, and its practitioners became the custodians of real time. Although economic history had its immediate academic origins in the late-nineteenth-century historicist reaction to the increasingly abstract nature of economics, it used an historical–inductive methodology that can be traced back into the ancient past. As shown in Chapter 3, while the methodological lineage of modern economic history can be traced back through the millennia, that for deductive economics can only be followed through the centuries. Hence it is wrong to suggest, as many have done, that this approach to reality is the progeny of economics and history – it is, in fact, the older sibling of both.

The institutional form assumed by the discipline has varied throughout the English-speaking world. In Britain, economic history quickly achieved separate departmental status within the university system, whereas in North America and Australasia, separate status was either never achieved – as in the USA – or was delayed until the university boom in the 1960s – as in Australia. The major determinant of independent status, it is argued here, is the degree to which the economics profession has taken an abstract–deductive stance. For example, in Britain, economics became highly abstract and non-empirical during the closing decades of the nineteenth century, whereas in regions of recent European settlement – USA, Canada, Australia, and New Zealand – economics, which reflected different economic circumstances, tended to be more pragmatically empirical. As the focus of economics in 'new' countries was more applied than in Britain, economic history was able to play a greater role in economic studies. This situation remained largely unchanged until the 1960s despite the increasingly deductive approach taken by economics, particularly in the USA, largely because of the establishment of institutional traditions. The university boom of the 1960s, which had the potential to change the institutional traditions, had a differential impact upon the organisation of economic history in 'new' countries. In the USA this boom coincided with the cliometric movement which, as it was fuelled mainly by economics graduates, did not lead to separate departments of economic history; whereas in Australia, the sudden demand for economic history was supplied by an influx of economic and social historians from Britain, who demanded and, were willingly granted, separate departments of economic history.

In this chapter, which surveys the emergence of the intellectual discipline of economic history in the English-speaking world, attention is mainly focused upon the directions set by the pioneers of those traditions rather than the detailed work of their followers. The objective is to discuss the nature of the profession's three main traditions in the United Kingdom, North America, and Australasia, in order to determine what role they have played, and whether one or more of these traditions could bring real time back to economics. These traditions include British economic and social history, American 'cliometrics', and Australian analytical economic history. Although these three traditions developed in different countries during the late nineteenth century and the first half of the twentieth, by the late 1960s they had spread throughout the English-speaking world. Hence I will concentrate upon the emergent period in each case.

Because aspects of this story are well known, our focus here is upon issues that have been overlooked or misinterpreted. In particular, greater attention is given to the pioneering Australian tradition in analytical and quantitative economic history which, although extending back to the 1880s, has been entirely ignored in recent historiographies. And the discussion of the British and American approaches is largely concerned with possible new angles. In all cases, every attempt has been made to allow the pioneers of each tradition to speak for themselves.

THE BRITISH ECONOMIC AND SOCIAL TRADITION

In searching for the beginnings of the British tradition of economic and social history, some have suggested that the 'Scottish enlightenment' of the late seventeenth century – consisting of Adam Smith, David Hume, John Millar and James Steuart – were the 'progenitors of what we now call economic history'.[1] But the case is not particularly persuasive. These writers did employ historical material, but largely to 'illustrate' their economic and philosophical arguments which, to one of the real founding fathers, was an approach that amounted to the 'perversion' of economic history. Also, there was no continuity between their work and the final emergence of an identifiable intellectual discipline some two hundred years later. There can be little doubt that the effective beginning of British economic history must be sought during the early-to-mid-1880s in Oxford and Cambridge.

The pioneers – Toynbee and Cunningham

Formal courses in economic history were initiated at both Cambridge and Oxford in the 1880s. This was a predictable outcome of an academic structure in which, at both ancient centres of learning, political economy was taught under the aegis of history, and of a time when the influence of the English historical school was at its height. At Oxford, economic history was introduced in 1881–82 by Arnold Toynbee, a few years after his appointment (1878) as tutor in political economy at Balliol College; and at Cambridge, William Cunningham introduced a separate paper on economic history in the History Tripos in 1885, some eighteen years before Alfred Marshall successfully established his Economics Tripos.[2] While these formal courses in economic history were taught under the academic banner of history, deductive theory was acknowledged to a greater extent by the founding fathers (particularly by Toynbee) than by those who followed them.

Arnold Toynbee

Arnold Toynbee's interest in economic history was stimulated by 'the social problems which are agitating the world now'. He argued that economic historians 'must pursue facts for their own sake, but penetrated with a vivid sense of the problems of your own time. This is not a principle of perversion, but a principle of selection'.[3] And, in pursuing the history of current problems, the economic historian should employ both induction and deduction, while the 'right method in any particular case must be largely determined by the nature of the problem'.[4] Further, he argued that:

> History . . . is better understood when studied in connection with Political Economy; for the latter not only teaches us in reading History to look out for the right kind of facts, but enables us to explain many phenomena . . . which without its assistance would remain unintelligible.[5]

While Toynbee thought that the English historical school correctly emphasised the strengths of the historical method – particularly by demonstrating not only the role of institutions in economics, but also those economic principles that are universal and those that are relative to time and place – he felt that they went too far when they completely rejected deductive theory in favour of the historical method: 'There is no real opposition between the two', he wrote. But he was aware that the deductive method could be misused by failing to examine assumptions closely and to bring the theoretical conclusions 'to the test of fact'; by employing unrealistic premises; and by 'the failure to combine induction with deduction'. 'But', he said wisely, 'this misuse of the method does not imply any radical faultiness in it'.[6] Toynbee also made quite clear his doubts that the historical method could 'by itself' determine the 'laws of economic progress':

> The historical method, on the other hand, is impotent of itself to give us a law of progress, because so many of the facts on which it relies are, in Economics, concealed from us. By the historical method we mean the actual observation of the course of economic history, and the deduction from it of laws of economic progress; and this method, while most useful in checking the results of deduction is, by itself, full of danger from its tendency to set up imperfect generalisations.[7]

One cannot but be impressed by the balanced viewpoint of Arnold Toynbee. The British tradition in economic history could not have had a better founding father. It is a tragedy that Toynbee did not have many real followers when economic history developed as a separate discipline in Britain. His most able pupil, W. J. Ashley, was initially convinced by the extreme argument of the historical school, and even when the historicist fire died within him, he rejected theory in favour of institutionalism. Possibly the only idea taken from Toynbee by the economic and social history tradition was that deductive theory is relative to time and place. Even today, the pre-modern branch of this tradition feels justified in rejecting modern economic theory. Finally, it is fascinating to notice Toynbee's distinction between the methodological principles of 'selection' and 'perversion', which Cunningham was later to adopt and use as a weapon against his colleague Alfred Marshall.

William Cunningham

William Cunningham's methodological discussion in *Industry and Commerce* (1882) is less sophisticated than that of Toynbee. Although Cunningham's introduction is lengthy, there is no clear discussion of the relative methods of induction and deduction. Instead, he stressed the interconnection between economic, political, social, and moral facts and ideas, and declared that 'politics are more important than economics in English history'.[8] Essentially he adopted a non-economic approach to economic history, by rejecting both an economic framework for his history of England between the 1080s and 1880s, and an economic analysis in favour of a 'comparative method'. In his own words:

> Economic students who attempt to investigate some remote epoch by the same analytic method – which they habitually apply to the economic phenomena of the present – may be led to assign an undue importance to one particular condition . . . We can only avoid those dangers by endeavouring to pursue an opposite method; we may begin with the political and social environment with the influences which are plainly observable in literature and public life; and working from a knowledge of the environment and of the forces actually in operation at any given time, we shall find how far the reported facts about buying and selling, meat and clothing, working and recreation become intelligible . . . We shall be better able to guard against the danger of generalising hastily from a few particulars.[9]

Therefore, as 'the growth of industry and commerce is so directly dependent on the framework of society at any one time, it may be most convenient to take periods which are marked out by political and social rather than by economic changes'.[10]

Cunningham claimed that while it was feasible to employ market-oriented economic theory in an analysis of modern England because of the prevailing, and compatible, policy of *laissez-faire*, it was not valid for other times and places because of extensive and incompatible government intervention. He argued:

> Economic forces in modern days may be treated abstractly and regarded as exhibiting the mechanical play of self-interest of individuals; while the government seems to have no part, but to preserve such security and order that this may operate freely. But History must trace out the conscious efforts, which were made from time to time, to develope [sic] the resources and expand the commerce of the realm; such deliberate endeavours were made through political institutions for political objects, and affected our progress for good or for evil.[11]

To Cunningham, political rather than economic considerations were the driving force in English history. He does not stop to consider what these forces are or whether they in turn result from economic motivation. But having rejected the tools of economics he is left only with the comparative method: 'By contrasting institutions or customs differing in time and place but with a strong superficial resemblance, we may learn to understand the true character of each'.[12] He is less than convinced, however, that the comparative method of the historical school will 'give us valuable generalisations or sociological laws'. Knowledge of the particular was enough for Cunningham.

In view of the non-economic approach adopted by Cunningham in his own work, it is little wonder that he was unable to accept either Marshall's approach to the past, or deductive theory with its underlying assumption of economic rationality. His 1892 outburst in the *Economic Journal* on these issues probably reflected a decade or so of frustration in attempting to maintain his non-economic methods in the face of Marshall's more analytical approach to the past. As demonstrated in Chapter 2, Marshall was caught out by Cunningham, not over the validity of employing deductive theory as a tool of analysis, but for the superficiality of his historical reconstruction.

The second generation – Ashley and Clapham

W. J. Ashley

Ashley's influence on the development of the British tradition was limited by the many directions taken by his academic career. He was initially influenced by William Stubbs, who in the late 1870s taught him that 'the roots of the present lie deep in the past',[13] then by Arnold Toynbee, who in the early 1880s convinced him that history was driven by economic forces, and finally by the German historical school, which later in the decade convinced him that economics must come from within history. While briefly under the influence of the historicists Ashley held that deductive thought could be replaced by inductive generalisations, but by the time he had published *English Economic History* (1888), dedicated to Arnold Toynbee, he was more concerned to assist in the development of a literature in economic history than to provide a basis for a body of inductive principles in economics. No doubt this reflected both the immediate demands of his academic career in economic history and the overwhelming complexity of the historicist exercise.

Unlike the economic histories of Toynbee and Cunningham, Ashley's *English Economic History* gave little space to a discussion of method, possibly because the phase had passed when there was a need to justify a new work in economic history. In the preface, however, he does emphasise that 'political economy is not a body of absolutely true doctrine . . . but a number of more or less valuable theories and generalisations'; that economic ideas have always existed but are rooted in the conditions of their time; and that 'modern economic theories, therefore, are not universally true; they are true neither for the past, when the conditions they postulate did not exist, nor for the future, when, unless society becomes stationary, the conditions will have changed'. Probably the main reason that Ashley and other historicists regarded economic theory as inapplicable in the past was their rejection of the possibility of universal economic man. Because of this relative view of economics, Ashley, like Cunningham and Toynbee before him, devoted substantial space to a discussion of contemporary ideas.[14] But while these ideas probably helped to illuminate the facts of the past, they were of no assistance in undertaking a formal economic analysis of the period of history from which they arose. It is hardly surprising, therefore, that the economic historians who followed focused, in the main, on economic circumstances, and left contemporary economic ideas to the historians of thought.

Also, there is little discussion about how he intended to tackle the task of writing history. Ashley provides only a general discussion of the two ways of proceeding – the use of a pre-industrial Ricardian deductive economics as defended by J. S. Mill and J. E. Cairnes, or the use of the historical method. Within the historical school there were two groups: those that 'try to free their minds at the outset of all *a priori* theories, and to see things as they actually are and have been, using deductive reasoning only as an occasional help in interpreting the results of their investigation', and those of '"the historical school" in the strict sense of the word' who 'attempt to discover . . . the laws of social development' in the belief 'that knowledge like this will not only give them an insight into the past, but will enable them the better to understand the difficulties of the present'.[15] Although he does not commit himself in the preface, it is clear from the text that he placed himself in the former category – that of economic historian rather than historical economist 'in the strict sense of the word'.

Not surprisingly, Ashley was more explicit about methodology in his 1893 inaugural lecture at Harvard. When addressing the question: 'What is economic history? And why study it?', he suggests, first, that some knowledge of theory is useful, not as a basis for formally analysing the past, but to suggest relationships that could otherwise be overlooked when studying documents which provide 'large and unknown stores of economical instruction'; secondly, that a study of documents is more useful than a study of statistics because it informs us about institutions – hence 'the infinitely greater importance of the first one hundred pages of Mr Seebohm's work [*Village Community*] than of all Thorold Rogers's voluminous collections'; thirdly, that it should be concerned with pursuing current economic problems back into the past, as his mentor Toynbee had done; and finally, that we might 'arrive at a more satisfying and intelligible conception of human society'.[16] Unfortunately, such a conception was denied him, largely because he, like those that followed him, refused to employ an analytical approach.

Ashley's influence in the development of the British tradition is not proportional to the amount of time he spent as a specialist in economic history. His work in the field only spanned the period from the mid-1880s to 1901; but in that time he was a tutor at Oxford, occupied the Chair of Political Economy at Toronto (1888–92), and the Chair of Economic History at Harvard (1892–1901). In the process he helped to transplant the British tradition to both Canada and the USA. Interestingly, he was also invited to take the Chair of History at the University of Sydney in 1890.[17] The Australian offer may well have arisen from the publication by Ashley of a six-part series of articles on economics in *The Australian Economist* in 1889.[18] After 1901 Ashley abandoned economic history for an institutionalist approach to the economy in his capacity as Head of the Faculty of Commerce at the University of Birmingham, England.

J. H. Clapham

The other major second-generation figure in the British tradition is J. H. Clapham, who learnt his economic history from Cunningham and his economics from Marshall.[19] Marshall taught him about the central role of economic man in society, mainly within the context of neoclassical microeconomics, and Cunningham impressed upon him that the materialistic drive is subsumed within the 'higher' moral, cultural, political, and social motives of mankind. On the one hand he learnt about the centrality of economic markets and institutions, and on the other of how these were subordinated to political structures and political forces. In *Principles of Economics*, Marshall, despite his habitual attempt to keep all disputants happy, says:

> the steadiest motive to ordinary business work is the desire for the pay which is the material reward of work. The pay may be on its way to be spent selfishly or unselfishly, for noble or base ends; and here the variety of human nature comes into play. But the motive is supplied by a definite amount of money: and it is this definite and exact money measurement of the steadiest motives in business life, which has enabled economics far to outrun every other branch of the study of man.[20]

Contradicting this position is Cunningham's belief in the importance of 'treating the history of economic phenomena in the past as wholly dependent on politics and political changes'.[21] At a later stage (1892), as we have seen, Cunningham explicitly rejected the concept of economic man. Students such as Clapham, therefore, were being presented with two incompatible views of the world, one dominated by economic man and the other by political man. But while Marshall was at pains to define and explore *homo economicus*, Cunningham fails to explain the nature of political man.

A student who took both masters seriously would have run the serious risk of being afflicted with intellectual schizophrenia. That Clapham did take both seriously may account for the ambivalent attitude he had to economic history. The first volume of his *Economic History of Modern Britain*, which is dedicated to both Marshall and Cunningham, indicates his high regard for both teachers. In the preface he explained: 'I have ventured . . . to dedicate this volume to the memory of two Cambridge economists of very different temper and outlook. One of them [Marshall] told me, twenty-five years ago, that it was my business to write something of the sort. The other first taught me Economic History.'[22] Loyalty to two masters, who saw the world in entirely opposite ways, probably contributed to his negative and self-deprecating view of economic history. In Chapter 2, we saw that Clapham in his 1929 inaugural lecture deferred to the rest of the history profession when he said: 'I am no rival. Yours is the higher work. I want to help.'[23] Twenty years later, when his last book, *A Concise Economic History of Britain*, was published posthumously, he had not changed his view. In the preface he wrote: 'Of all varieties of history the economic is the most fundamental. Not the most important: foundations exist to carry better things . . . Economic advance is not the same as human progress.'[24] Clapham appeared to believe that although economic forces determined the foundations of society, the social and political superstructure was fashioned by separate and higher considerations and motives, and was the province of the more prestigious branches of the history profession. Clearly, he rejected Cunningham's non-economic approach to *economic* history, but as if to atone, he claimed that economic history was not as important as political history. The militant historicist had left his mark.

Clearly, Clapham's work is informed by theory, and from time to time there are self-conscious references to the interests of 'the economist', but it does not involve explicit economic analysis. And while he uses official statistics on outputs, inputs, and factor returns, he does not construct and employ macroeconomic aggregates such as national income, investment, or money supply in his work. He is not, therefore, an analytical and quantitative economic historian. In his main work, *The Economic History of Modern Britain* (1926–38), Clapham describes the economic organisation and condition of Britain, and the way in which this changed during the century leading up to the First World War. Even the structure of the various volumes in this series owes little to economics, focusing as it does on the main productive sectors of the economy – agriculture, industry, commerce – the economic activities of the state, and the resulting conditions of 'life and labour'. Even his last book, *A Concise Economic History of Britain*, employed a similar non-economic, largely pragmatic, structure. The latter work is, indeed, rather disappointing, both in terms of its failure to employ, in any analytical sense, major statistical sources such as Domesday Book or the political arithmetic of Gregory King and his followers, and its failure to provide any sense of what was driving the massive longrun changes in British society and its standards of living. Instead, we are presented with a non-analytical description of the changing economic characteristics of British society during the very longrun. Finally, even his specialist history, *The Bank of England* (1943 and 1944), is written in a chronological and general way, showing few signs of an analytical or quantitative approach. None of his work breaks out of the framework of the British tradition established by Cunningham.

Epilogue

It is difficult to say why one of several pioneering approaches to a discipline prospers. The fact is that Cunningham rather than Toynbee appears to have been the model adopted by subsequent scholars in the British tradition. This probably had something to do with Toynbee's premature death, and to Cunningham's long and vigorous advocacy of economic history as a separate discipline.[25] Also, the conflict within economics between the historical and deductive schools, which led to the emergence of economic history as a new entity, produced a mutual antipathy between deductive economists and the descendants of the historicists. Accordingly, when economic history began expanding in British universities after the First World War it was supplied with recruits from history rather than from economics. This in turn led to an even greater divergence between the two disciplines, with the economic historians feeling more comfortable with other disciplines in the social sciences such as sociology. This arrangement appears to have been formalised by the recent (February 1991) addition of a new qualifying subtitle – '*a journal of economic and social history*' – to the Economic History Review. I am not, of course, including in this tradition those who have been sympathetic to the 'new' economic history, or indeed those of an even younger generation who have embraced it warmly.

Not only did economic history become isolated from economics, it also fragmented into distinct periods – modern, early modern, later Middle Ages, Anglo-Saxon and so on – that were stoutly defended against outsiders. This contrasted with the pioneering contributions of Cunningham, Ashley, and Clapham, which ranged over the very longrun. This was probably a natural outcome of the expansion that occurred in the profession after the Second World War, but it gave a fragmented view of British economic development that few – a notable exception being M. M. Postan – could, or were permitted, to transcend. Interestingly, N. S. B. Gras predicted this problem as early as 1920. While he thought that economic history (in the USA) should develop as a separate discipline, he warned of the dangers of erecting barriers around time periods. He acknowledged that some of the best scholars had used 'periods', rather than his preferred 'topics', as a device in writing economic history, but was concerned that: 'In the hands of lesser scholars it leads to the erection of stone walls at the beginning and at the end of the periods in question.'[26] In his opinion: 'By the principle of development or evolution, alone, can researches in the natural or social sciences be summed up'. Even today these barriers that Gras warned us about seventy years ago are still in place, particularly between the later Middle Ages and both the early modern and Anglo-Saxon periods. While the period since 1500, and more particularly since 1700, has seen the introduction of new ideas and methods, the economic and social historians of the medieval era are still rebuilding the stone walls each time one is breached by new ideas. This has delayed the attempt to take up the challenge of the pioneers of the British tradition to see history as a continuous process.

THE AMERICAN CLIOMETRIC CONTRIBUTION

The early years

It is not intended to provide a comprehensive outline of the development of economic history in North America; merely to sketch a background against which cliometrics emerged.[27] While it is interesting, as some have done, to seek early examples of works that are both economic and historical, it is clear that economic history in North America had the same source as the British tradition – the German and English historical schools. But in contrast to British experience, economic history in the USA did not become established as a separate discipline, or as a clearly defined sub-group in the disciplines of either economics or history. Prior to the Second World War, economic history was largely the by-product of the research of either economists or historians.

Writing in 1920, Gras said: 'economic history, until now in the able hands of historians and economists, might henceforth more profitably be lodged with specialists in the field'.[28] But eleven years later he was unable to record any change. 'Economic history in America', Gras claimed, 'has had no leader to produce a monumental work . . . The universities have generally neglected the study of economic history . . . There has been a lack of controversy, even of intellectual resilience, in the field.'[29] As will be shown, apart from a brief, and largely unsuccessful, flourishing of staple-thesis history at Toronto in the 1930s and 1940s, and of entrepreneurial history at Harvard in the early 1950s, it was not until the emergence of the cliometric movement in the late 1950s and early 1960s that North America produced a distinctive school of economic history.

This contrast with British and, as we shall see, Australian experience, is stark. It has been explained largely in terms of the empirical and institutional nature of American economics prior to the First World War. American economists of the early twentieth century, according to A. W. Coats, regarded 'departments of economic history as unnecessary because so many of them already took account of the historical dimensions of their subject in their teaching and research'.[30] This closer link between economics and history owed much to the pragmatic institutionalism of American economics, which was strongly influenced by the German historical school and which attempted to grapple with the distinctly different economic circumstances of a region of recent settlement. It was the pragmatically empirical approach that gave rise to the National Bureau of Economic Research (NBER) in 1920 under the guidance of the institutionalist W. Mitchell – who was inspired by T. B. Veblen (1857–1929), that eccentric critic of deductive economics[31] – and the economic historian E. F. Gay.

There are interesting parallels with the institutional–empirical approach adopted by economists in Australia, another region of recent white settlement which also pursued a more pragmatic approach that conflicted with British deductive economics.[32] But, curiously, North America did not produce an analytical and quantitative tradition in economic history to compare with the Coghlan tradition in Australia. It should be made clear that the national accounting work of Simon Kuznets in the inter-war period, which was similar to that of Colin Clark in the UK and Australia at the same time, was not part of any American tradition in economic history. It was the sort of pragmatic response to developmental conditions in 'new' regions that found a comfortable home in institutional economics. As will be shown, the pioneering work in modern national accounting was undertaken by T. A. Coghlan in Australia in the 1880s, and was merely reinvented by Kuznets and Clark in the 1930s.

In searching for the origins of American economic history, Gras traced individual works in what he called 'commercial history' back to the late eighteenth century.[33] But such works were few and far between before the 1880s, and cannot be regarded as part of a self-conscious intellectual discipline of economic history. Gras admits that: 'It would require some stretch of imagination to see in all this the effective beginning of economic history.'[34] The first significant work was by A. S. Bolles in 1878, but this was an industrial history written by an economist rather than an economic historian.[35] Even in the universities the main contributions to economic history were made by historians – such as Bancroft, Turner, McMaster, Channing, Phillips, Buck, Frank, Westerfield, Rostovtzeff, Gray, Cheyney, Dietz, H. B. Adams; and economists – such as H. C. Adams, Taussig, L. C. Marshall, Mitchell, Dewey, Wright, Bullock, Ripley, E. Jones, Dowrie, Kemmerer and Ely. Only Harvard made any real attempt to attract specialist economic historians, such as S. Dunbar (from 1883), W. J. Ashley (1892–1901), E. F. Gay (from 1902), N. S. B. Gras (from 1927) and, later, A. Gerschenkron. Even Columbia and Yale, which established chairs in economic history after the turn of the century, made a practice of filling them with non-specialists. It is interesting that, during the period before the Second World War, the recognised doyen of economic history in the USA, Edwin Gay, published very little apart from a few articles on enclosures in the English publication *Transactions*. His contribution was largely restricted to the training of the next generation of economic historians, basically in the historicist tradition with influences from both the German and English historical schools. Gras claimed in 1927 that Edwin Gay 'has come nearest to founding a school of economic history'.[36] The other North American figure who came close to founding a school in this period was H. A. Innis at Toronto, with his basically historical work in the 1930s and 1940s on the staple approach to Canadian economic development.[37] But it was an approach that had few successful imitators.

Generally speaking, American economic history research in the pre–Second World War years was not closely associated with mainstream issues in economics – as it was in Australia – such as economic growth, distribution, and welfare. Instead, it was associated with institutional sideshows, such as business studies, agricultural economics, industrial relations, and banking. This fragmentation is reflected in the avenues available in the USA for the publication of economic history articles. Before the *Journal of Economic History* was established in 1941, relatively late in comparison with the English *Economic History Review* (1927), a number of specialist journals carried articles in economic history. These journals included *Agricultural History* (1926) and the *Journal of Economic and Business History* (1928). But in contrast to their transatlantic cousins, American economic historians were working more closely with economists, at least of the institutional variety, and were more comfortable with quantitative data. And while Gras could conclude in the late 1920s that economic history in America 'had no intellectual resilience' and that 'there has never been published a single outstanding general treatise on American economic history',[38] it was a far more fertile ground than existed in England for the reception of new developments in economic history, such as entrepreneurial history in the early 1950s and cliometrics in the late 1950s and early 1960s.

Between the Second World War and the late 1950s, economic history in North America did gain specialised professional status, symbolised by the establishment of the Economic History Association and the establishment of the *Journal of Economic History*. But this separate identity was achieved by economic history moving away from economics – which was becoming increasingly deductive in orientation – and towards history, just as had happened in Britain after the First World War. In 1960, on the eve of the emergence of cliometrics, one commentator lamented the growing 'alienation' between economic history and economics, which he thought had:

> proceeded far beyond the normal differences of emphasis that might be expected to exist between specialists in the inductive and deductive methods. Economic theory today, in most of its branches, neither draws on economic history for its data nor goes to economic history for empirical verification. Economic history, for its part, commonly uses only the crudest of the tools in the economist's tool-box, and displays almost complete indifference to the refinements in analytical methods that occupy the theorist's working time.[39]

Instead, economic history was seen to have become part of general history where it achieved 'high prestige' but at the cost of becoming 'descriptive or institutional economic history'.

Even the much-trumpeted entrepreneurial history group at Harvard in the early 1950s did not improve the discipline's standing in the eyes of economists, despite the blessing given to it by Joseph Schumpeter. This first US initiative in economic history was not to be as successful as the second, which occurred a decade later. While the Research Centre for Entrepreneurial History, which was established in 1948 by Arthur Cole, was innovative, it made the mistake, at least as far as economists and economically trained historians were concerned, of taking its interdisciplinary approach too far. After the death of Schumpeter in 1950, the entrepreneurial group came increasingly under the influence of sociologists, with the result that analytical economic historians such as the Australian, Noel Butlin, became disillusioned and left Harvard, complaining that 'the Entrepreneurial History Centre came to be dominated by sociologists . . . I came back to Australia to work on the [whole] economy'.[40] Others, such as Hugh Aitken, found the 'lack of interest shown by economists' in this experimental project to be 'distressing and discouraging'.[41] Where this project probably went wrong was to focus upon a single agent in the process of economic development and to abstract it from the context of the economy as a whole. This, of course, went beyond the interests of economists, who take the motivation of economic agents as given. But it did give rise in 1948 to the journal *Explorations in Entrepreneurial History* which later (1969) changed its name to *Explorations in Economic History*, and, interestingly, became the journal of the second major American contribution to economic history – cliometrics. At the end of the 1950s, therefore, one observer, Hugh Aitken, saw economic history in the USA as leading a 'precarious existence' and called for a closer relationship with economics. Indeed, he felt that economic history in Canada, mainly because of the role of Harold Innis at Toronto, had 'a more influential and respected position in the academic world than it does in the United States'.[42] Aitken incorrectly saw Innis' work on staples bringing economic history closer to economics because of the latter's passing interest in development issues. While Innis' work was an interesting example of inductive generalisation, it is only concerned with the process by which an expanding export staple impacts upon the rest of the economy, and has little to say about broader issues of growth, about inflows of capital and labour, and about the way in which growth is transferred to the urban sector. Clearly, Aitken had little premonition of what was about to occur in the USA, and which would spread throughout the world.

Cliometrics

The cliometrics phenomenon (it was not a revolution as some have claimed) took many economic historians, even close observers of the American scene, by surprise. The romanticised account of this event is that it occurred with a great burst of energy, enthusiasm and bombast in the late 1950s and early 1960s, beginning at Harvard and quickly transferring to Purdue where the first cliometric (although not yet known by this name) meeting occurred in December 1960. But, in fact, the emergence was more gradual, less dramatic, and even more modest than this interpretation allows. While there was enthusiasm there was also self-doubt, and much of the agitation was generated by those who opposed this 'new' way of writing history.

The essence of cliometrics, as others have already emphasised,[43] is the use of economic theory and quantitative techniques in reconstructing the past. As it has turned out, the theory is more important than the quantitative techniques, and the range of the quantitative techniques is wide and depends upon the issue under examination. In fact, regression analysis is not essential to cliometric work, despite the assumption to the contrary made by general historians. As we shall see, the pioneers of cliometrics – A. H. Conrad and J. R. Meyer, and Robert Fogel – made only limited use of sophisticated statistical techniques. This realisation has encouraged some practitioners to seek a more suitable name than either cliometrics or, the earlier choice, econometric history. Donald McCloskey, for example, has opted for the even more misleading term 'historical economics' – misleading because it has long been used to refer to the historical school of economics in the nineteenth century, and because that school was diametrically opposed to the use of deductive theory embraced by cliometrics. But the latter helps to explain the passionate resistance to cliometrics exhibited by scholars in the British tradition of economic and social history – the descendants of the early historicists. Indeed, McCloskey is over-optimistic when he suggests that 'the victims of the [cliometric] invasion have become tolerant of economic standards of logic'.[44] It may be true that traditional economic historians interested in the modern period have withdrawn from open hostilities, but those who work on the medieval period are still fighting a strenuous rearguard action.

An interesting question is: why did the American brand of cliometrics have such rapid and impressive success in North America and other parts of the world? In 1985, for example, Northwestern University played host to the First International Cliometrics Conference – which marked the twenty-fifth anniversary of the cliometrics group – with participants from around the world.[45] This apparently unheralded emergence of cliometrics is explained by A. W. Coats, a less than enthusiastic observer, in the following terms:

> The short-term success of the cliometric movement was facilitated by certain enduring characteristics of American intellectual life . . . and by the structure and functioning of the higher educational system . . . The openness to new ideas; the high premium on novelty; the concomitant tendency to exaggerate claims to originality; a disrespect for past achievements . . . the overinvestment of resources in academic trivia and gimmickry; the academic pressure to 'publish or perish' and . . . the veritable passion for quantification and measurable standards of performance.[46]

This explanation is cast largely in social and cultural terms, whereas a more persuasive explanation has economic foundations. First, the reported 'characteristics of American intellectual life', which Coats appears to view as cultural defects, appear to be a function of a society that is more open to competitive forces than British society, which is his unstated benchmark for comparative purposes. Secondly, the *strength* of the cliometric surge owed much to its timing, coinciding as it did with the rapid expansion in the supply of PhDs in economics during the 1960s, and their reduction after the early 1970s.[47] It was, in other words, largely a function of demand and supply in the academic labour market. The end of the boom in cliometrics (measured in terms of the rate of growth of new entrants and publications), but not the end of its influence, had much to do with the end of the wider academic boom in the USA. Coats appears to confuse the end of the rapid growth of cliometrics with the end of its influence when he refers to the 'short-term success of the cliometric movement'. The impact of cliometrics upon the economic history discipline is permanent; the change is irreversible; economic history will never be seen by the world in the same way again. Thirdly, the emergence of cliometrics, rather than the speed at which this occurred, is not so surprising. The ground in America had been prepared: economic history had had a long association with a pragmatic brand of institutional economics that had given rise to the work of Simon Kuznets (the PhD supervisor of Robert Fogel), and it was comfortable with quantitative as well as qualitative data. And there was a pre-existing tradition in analytical and quantitative economic history in Australia that had emerged from similar conditions about eighty years previously. Cliometrics was, in effect, a rediscovery and popularisation of this earlier Antipodean tradition. It is not about to disappear.

The initial cliometrics manifesto was Conrad and Meyer's *The Economics of Slavery and Other Studies in Econometric History*, published in 1964, although some of the chapters had appeared in preliminary form elsewhere between 1957 and 1961. They stated their intention quite specifically, but quite modestly – because they wanted to 'avoid *methodenstreit* . . . arguments about precedence or relative importance'[48] – as follows: 'We had been talking and arguing about evidence, explanation, and proof in history for several years, trying to find a link between the new techniques of econometric and numerical analysis that we were learning and the problems of explanation in the history we were reading.'[49] This focus led them to devote the first two chapters of their book to the use of theory and statistical inference in economic history. They explained: 'This chapter is an attempt to examine critically the function of theory in historical research and particularly in economic history.'[50] Their work was presented as a tentative voyage of discovery, which began under the critical eye of their mentor, Alexander Gerschenkron. There is little here that is dogmatic or bombastic, and while their use of theory is central, their use of statistical technique is basic and sparing. There is little here, in other words, that could be regarded as provocative. Had one been unaware of the anti-deductive, anti-economic rationality of the descendants of the historicists, one would have been left wondering what all the fuss created by the British tradition was really about. Surely youth, enthusiasm, and excitement about new interpretations deserved a more generous reception.

Robert Fogel's book, *Railroads and American Economic Growth*, also appeared in 1964, and it too made relatively modest claims for the approach adopted. Fogel explained that:

> The pages that follow contain a critical evaluation of the proposition that railroads were indispensible to American economic growth during the nineteenth century. The analysis is, for the most part, based on the accepted narrative of the American transportation system. However, certain limited but crucial aspects of the narrative are re-examined with the aid of the analytical and quantitative techniques of modern economics.[51]

These points are elaborated further: 'I do not mean to suggest that the highly detailed, explicitly theoretical, and quantitative type of analysis presented herein is the only useful type of economic history'; and again he claims that the book attempts to 'emulate that blend of the empirical and deductive which is the hallmark of his [Simon Kuznets'] studies'.[52] Although Fogel claimed to have reinterpreted the role of railways in American growth, he was modest, although confident, about the techniques he employed. Basically, Fogel uses economic theory to set up the issue of whether railroads were indispensable to American growth, and then employs some simple arithmetic to make his point. As McCloskey reminds us, there is no sophisticated use of statistical analysis in this book.[53] Anyone accustomed to logical thought could follow Fogel's arguments.

Admittedly, there is more technical econometric work in Volume 2 of *Time on the Cross*, published by Robert Fogel and Stanley Engerman a decade later in 1974, but even here it is not overwhelming; and, in any case, the argument is presented in non-technical terms in the first volume. But in *Time on the Cross* Fogel had become much more assertive about his economic and statistical methods. Once again he claimed, more boldly than before, that he had overthrown the traditional interpretation. This, Fogel argued, had been achieved by the 'processing of large quantities of numerical data' made possible by 'a series of rapid advances [after the Second World War] in economics, statistics, and applied mathematics, together with the availability of high-speed computers'.[54] Actually, he overstressed the technical side of his work in this book, claiming that his reinterpretation 'hinges to a large extent on certain broad methodological questions, and particularly on the role of mathematics and statistics in historical analysis'.[55] Interestingly, a decade or so later, he played down the role of technical methods, when in an interview in 1990 he claimed (equally misleadingly):

> When I first tried to define what was novel about cliometrics, I mentioned the gathering of new evidence, but I made the more explicit use of theory predominant. I think I would now reverse that order . . . In a sense cliometrics has restored the old emphasis on archival work.[56]

The problem that historians may have had with Fogel's work is not with the technical difficulties of his empirical methods, but with the uncompromising economic logic of his conclusions and interpretations, which cut across the fashionable historical myths. In some quarters, opposition to the cliometricians was a reincarnation of the militant historicist attack upon economic rationality and deductive economic thought, directed this time not at the economics profession – which had long since ceased to notice or care – but against this American version of analytical economic history.

Donald McCloskey, who has become the main apologist for cliometrics, also places greater emphasis in his own work on the use of theory than upon the use of statistical technique. In *Economic Maturity* (1973) he makes little of the technical aspects of his approach – perhaps reflecting the moderating influence of Alexander Gerschenkron, his supervisor – claiming instead that the 'methods are eclectic and the pieces of evidence varied'.[57] The mathematics in this work, used to formalise his economic thinking, have been confined largely to footnotes, and the quantitative techniques, used to 'illustrate' his deductive arguments, get little beyond simple arithmetic. McCloskey's approach to history is basically deductive rather than inductive and, accordingly, he has little interest in econometrics, which may explain his curious attack upon statistical techniques[58] and his willingness to include economists who dabble in history (and court the wrath of Cunningham) in his category of 'historical economists' along with more mainstream, analytical economic historians.[59]

An interesting example of the logical extension of McCloskey's vision of economists' history is the 'general equilibrium history' pioneered so effectively by J.G. Williamson. An early example of this approach is his *Late Nineteenth Century American Development*, published in 1974, the same year as Fogel and Engerman's *Time on the Cross*. None of that restraint or caution shown by the pioneers of cliometrics, in relegating mathematical model-building to footnotes or appendicies, is apparent here. Williamson, a development economist as well as an economic historian, presents his history as a technical exercise in economic analysis. He begins with a very brief 'survey of the historical terrain', in which stylised facts and issues are reviewed; follows this with a detailed mathematical specification (72 equations) of a simple model of regional growth; and, finally, uses simulation techniques to compare the model's predictions with a reality summarised by time-series data relating to theoretically determined macroeconomic variables. There is a boldness here that would have unnerved even the pioneers of general equilibrium analysis.

In his opening sentence, Williamson informs the reader of what he/she might expect: 'This book is an economist's attempt to make sense out of a critical period in American history'. And in discussing his approach, Williamson informs us:

> Although the historical topics taken up in the pages following are conventional, the methods utilized are not, even for cliometricians. The counterfactual is used extensively, of course. But more to the

> point, the book relies very heavily on general equilibrium analysis. Furthermore, simulation is the quantitative mode which translates the formal model into a tool for historical analysis. Cliometricians have been reluctant to take this step in their work, perhaps because they are not sufficiently confident in the usefulness of general equilibrium models for empirical work in economic history.[60]

In contrast to Fogel's work, the emphasis here is more on the novelty of the method than on the interpretation, although new interpretations are certainly provided. Williamson views his task as a risky procedure, but is confident that the new insights it provides outweigh the limited focus on reality provided by such simplified general equilibrium models. Obviously, such a limited framework cannot encompass the complexity of historical reality, but Williamson's work in this and various other books has generated many fertile ideas. However, work of this nature raises the serious issue of communication with other social scientists, to say nothing of the educated lay person. Once again in contrast to Fogel, Williamson makes no concession to the non-economist, with the result that while Fogel's work has been widely read, Williamson's general equilibrium work has only been consulted by economists interested in historical application of their theory, and economic historians with a strong background in economics.

Possibly the most interesting and potentially useful attempt to use theory to explain history is *The Rise of the Western World* by D. North and R.P. Thomas. They employ neoclassical theory to 'explain' the changing economic systems of Western Europe from feudalism to capitalism. In discussing this attempt they say:

> We have developed a comprehensive analytical framework to examine and explain the rise of the Western World; a framework consistent with and complementary to standard neo-classical economic theory . . . This is an interpretive study . . . rather than an economic history in the traditional sense.[61]

The argument is cast in terms of evolving institutional arrangements that resulted from changing property rights and transaction costs as markets emerged during the period since AD 900. Once the 'analytical framework' is developed, they attempt to illustrate its plausibility by reference to both qualitative and quantitative evidence. On the quantitative side, this evidence includes longrun series of wages, prices, and rent that are available in the existing literature. Unlike

Fogel, they do not attempt to explore new sources of data. This approach differs from the other pioneers of the new American economic history because it is concerned not with the direct application of modern economic theory to the past, but with an examination of one of the *ceteris paribus* conditions of economics – that institutional arrangements are unchanged. Conrad and Meyer, Fogel and Engerman, and Williamson are all prepared to employ modern economic theory that is not concerned with real, or indeed analytical, time, whereas North and Thomas are not. As will be suggested in Part II of this book, much of the growth that occurred during the past millennium was due to causes that economic theory leaves out. Hence the approach of North and Thomas is potentially very important. But although they stress that their work is only an 'agenda for new research', little of this new research has been forthcoming. The difficulty that N. S. B. Gras noted – of transcending the stone-walling of period specialists – would appear to have barred the way. A modest attempt to hurdle these stone walls has been made in Chapters 5 to 7 in this book.

Cliometrics, the new American economic history, began in the late 1950s with economists tentatively using theory explicitly in the writing of history. Owing to the close association of economics and history in America, it was not surprising to see the cliometric pioneers drawing support from empirically-minded economists such as Simon Kuznets, and economically-minded historians such as Alexander Gerschenkron, for inspiration. There was no tradition of analytical economic history in the USA that they could look to. While some of the lesser figures in the cliometric movement, mainly young economists who were briefly attracted like moths to this new night light, may have made extravagant claims for the 'new' approach, the pioneers and other major figures were relatively restrained and modest. The almost hysterical reaction to this movement had little to do with the empirical techniques employed, as they were quite straightforward, and much to do with the rational economic approach that had long been rejected by the historicists and their descendants in the British tradition. What had been overlooked by both protagonists, however, was a long-established tradition in analytical economic history in Australia.

THE AUSTRALIAN ANALYTICAL TRADITION

The Australian, or Coghlan, tradition is the oldest continuous tradition in analytical and quantitative economic history in the

English-speaking world. In the light of this fact, the American cliometric movement should be seen as a 'rediscovery' of the methodology of this tradition, together with its world-wide transmission, rather than as an entirely new development. Of course, there are a number of isolated scholars elsewhere in the world who one may wish to include in the analytical economic history category, but in the main, either they were not part of a living intellectual movement, or they did not always regard themselves as economic historians. Individuals that do come to mind in Britain are Thorold Rogers in the second half of the nineteenth century; A.K. Cairncross, Brinley Thomas, and T.S. Ashton in the 1940s and 1950s; and Phyllis Deane and Charles Feinstein from the 1960s to the 1980s; and in the USA, possibly E.J. Hamilton in the 1930s and 1940s, and C.P. Kindleberger from the 1950s to the 1980s. But there was no continuous tradition here. Deane and Feinstein were not influenced by Rogers, but by economists such as Simon Kuznets. Also, it can be argued that many in this list regarded themselves, at least for a substantial part of their careers, as economists rather than economic historians. There were, of course, others in Britain who were sympathetic to cliometrics when it emerged, such as Alec Ford, Brian Mitchell, Patrick O'Brien, Roger Schofield, Barry Supple, and Tony Wrigley; a younger generation that embraced it warmly, including Nick Crafts, Roderick Floud, Robert Millward, and Nick von Tunzelmann; and now *their* students.

In Australia, however, a continuous intellectual tradition has existed in analytical economic history since the 1880s; a tradition of economic historians who employed economic theory and quantitative techniques, rather than of economists who dabbled in history. This Australian tradition began with the pioneering social accounts and quantitative essays of T.A. Coghlan from the 1880s to the early 1900s, and his analytical economic history of Australia in the 1910s; and included the work of Edward Shann in the 1910s to 1930s; the monetary and 'real' economic histories of the remarkable Butlin brothers, Sydney and Noel, from the 1950s to the 1980s; and the students of the Butlins – Alan Barnard, Colin Forster, Neville Cain, Boris Schedvin, Gus Sinclair, and Graeme Snooks – and the students of *their* students – Bob Jackson, Ian McLean, David Pope, and Greg Whitwell.[62] This group has been joined by other scholars influenced by the compatible American tradition in cliometrics including Gary Hawke, Steve Nicholas, Jonathan Pincus, and Ralph Shlomowitz. For over a hundred continuous years, scholars in the Coghlan tradition have employed economic theory and quantitative techniques to reconstruct the past, initially of Australia but more recently of Europe as well.[63] The quantitative armoury of this group was increased in the mid-1950s with the econometric history by E. Dunsdorfs (originally a student of E. F. Heckscher in Stockholm), entitled *The Australian Wheat-growing Industry, 1788–1948*. Dunsdorfs' statistical work, if not his theory, was more sophisticated than the early American cliometricians who came after him in the late 1950s and early 1960s. Hence the students of Noel Butlin were receptive to the American cliometric movement because it fitted neatly into their own tradition. The pioneering work of these Australian scholars, however, was overlooked in the cacophony of self-congratulation by enthusiastic young American scholars, and hostile rejection by conservatives in the British tradition, all of whom were blissfully unaware that the Americans had been anticipated by some seven decades. Not surprisingly, it took the 'reinvention' of the analytical economic history tradition by the world's leading power to transmit it throughout the world during a widespread boom in university education.

In institutional terms, the early Australian tradition evolved in departments of economics rather than history or economic history. The two key economics departments in this process were those at the University of Western Australia under Edward Shann between 1913 and 1934 (he began teaching economic history at the University of Melbourne as early as 1905 and at the University of Queensland from 1911 to 1913); and at the University of Sydney under R. C. Mills (who inspired the Butlin brothers with his scholarly but empirical and institutional approach to economics) in the 1930s. Interestingly, both Shann and Mills were products of the University of Melbourne and the LSE. After the Second World War the focus of the Australian tradition shifted to the Australian National University (ANU) and Sydney (until the shift of Syd Butlin to the ANU to join his brother in 1971) and flourished briefly (mid-1970s to late 1980s) at Flinders University in South Australia. The establishment of separate departments of economic history in most universities during the university boom of the 1960s had much to do with graduates of the British tradition migrating to Australia.[64]

Antecedents of the Coghlan tradition

While the Australian tradition effectively began with the quantitative work of Timothy Augustus Coghlan in the 1880s, there are a number of earlier writers, going back to the 1850s, who should be mentioned because of the intellectual climate that they helped to create. These include, possibly to the surprise of some, W.S. Jevons, that rare combination of empiricist and theorist, and David Syme, the newspaper-owner-cum-historical-economist.

W.S. Jevons lived in Australia for almost five years, between June 1854 and April 1859, where he was employed as an assayer in the Sydney Branch of the Royal Mint. These were important formative years for Jevons – from the age of nineteen to twenty-four – during which he began to formulate attitudes and ideas that were to make him the greatest economist of his generation. As J.M. Keynes said: 'His long period of solitary thought in Australia at an age when the powers of pure originality are at their highest, had been abundantly fruitful. For soon after his return, the outlines of his principal contributions to knowledge were firmly fixed in his mind.'[65] These were the years when his interests changed from natural science – his departure for Australia, appropriately for economic reasons, had disrupted a science degree course in London – to 'the scientific investigation of man' which 'is surely a work worth a lifetime, and one not excelled in interest by any other'.[66]

During these Sydney years the precocious intellectual abilities of Jevons were noted by his learned colonial contemporaries and became part of the general cultural environment. His writings at this stage were characterised by careful observation of natural and social phenomena, scrupulous recording of quantitative evidence, and a perceptive analysis of reality. While Jevons' earliest published work in Australia concerned his meteorological observations, it is his socio-economic writing that attracts our interest. In a series of articles on land and railway policy, and an interesting socio-economic survey of Sydney, Jevons demonstrated the analytical and quantitative ability that later was to attract attention in Britain also. His 1858 survey involves a methodical description of the living conditions of Sydney's population, in terms of streetscapes, housing types, commercial establishments, sanitary conditions, and occupations. This is of interest as it anticipates the early focus and work of T.A. Coghlan as Government Statistician for New South Wales (NSW) from 1886, and which would have been an influence upon him. Other matters that these two men had in common intellectually were an interest in urgent problems of the day, a detailed empirical approach to these problems, and a fine analytical grasp of the issues.

The role of David Syme in the development of the Anglo-Saxon form of historical economics has been discussed in Chapter 2. Our interest here is in the example he provided for the development of the Australian tradition.[67] Syme, as a historical economist, made no contribution to the methodological basis of the Australian tradition, which looked instead to deductive economics. Herein lies a fundamental contrast between the Australian approach and the British tradition, which was inspired by the historicists. That strain of economic and social history in contemporary Australia is not home-grown, rather it was imported from Britain after the Second World War, particularly during the 1960s and 1970s. What Syme did for the Australian tradition was to emphasise the need, for policy purposes, to treat Australia separately from Britain, and so emphasise that Australia had its own identity and its own objectives, and that these were to be pursued according to local dictates. This empirical/economic nationalism is a major characteristic of the Australian tradition as seen in the work of Coghlan, Shann, and the Butlin brothers, although they did not accept his historicist policy point of view.[68] The link is strong, because the nationalistic attitudes of Syme were widespread and well known, particularly as the 'father of protection'. Syme also emphasised the importance of economic development in 'new' countries, at a time when this issue was not fashionable amongst the neoclassicists in Britain. Growth and development have always been important issues in the Australian tradition irrespective of intellectual fashions in the economics profession in Britain.

T. A. Coghlan – the founder of the Australian tradition

Timothy Coghlan is one of the most remarkable figures in the entire international history of the discipline. None of the other economic historians of his time, from the 1880s to the 1920s, was as well equipped for the task. And there have been few since Coghlan's time that can stand against him. The tragedy is that his work, which focuses solely upon Australia, is so little known. While he is recognised in *The New Palgrave* with a short entry by Colin Clark, his work is no longer well known outside Australia. But this was not always the case: Coghlan's statistical work was highly regarded internationally in the 1890s; in 1893 he was elected Fellow of the Royal Statistical Society; and he was relied upon heavily by Mulhall in his popular *Dictionary of Statistics*. No one, no matter how innovative they are, who focuses solely upon Australian issues can hope to maintain more than a local reputation – contrast this with the reputations of those who have focused solely upon the UK or USA. Nor did it help his reputation that intellectually he was several generations ahead of his time. It is time, however, that Coghlan received international recognition for the enormous contribution he made.

Background

Even the bare outlines of Coghlan's life are impressive. He was born of Irish Catholic parents in Sydney in 1855, and entered the NSW public service in 1872, where he was appointed Assistant Engineer of Harbours and Rivers in 1881 and Government Statistician in 1886, a position he held until 1905. During these years, Coghlan was actively involved in a wide range of public service activities including being the Registrar of Friendly Societies; a leading member of official enquiries and Royal Commissions; and policy adviser to colonial governments on financial institutions, land and income tax, electoral reform, and local authority schemes. From his retirement in 1905 until his death in 1926, Coghlan lived in London where, for much of this period, he was NSW Agent-General, which allowed him to devote a large part of his time to writing his four-volume economic history of Australia, entitled *Labour and Industry in Australia*. His extensive contribution to public life was recognised in the award of the Imperial Service Order in 1903, a knighthood in 1914, and a KCMG in 1918.

But it is Coghlan's contribution to economic history that is the focus of our concern. It can be said without exaggeration that not until the 1960s could even a few economic historians around the world claim to possess the range and depth of skills deployed by Timothy Coghlan three generations before. During the first phase of his public service career, Coghlan, with a strong interest and ability in mathematics, became a qualified engineer. But, in a similar way to Jevons, he was to find the quantitative study of man in society more interesting and more significant than engineering and science, and in 1886 persuaded the colonial government that he was the person to occupy the new post of Government Statistician. As Statistician his mathematical and statistical talents had full rein. He transformed the statistical office from its traditional role of merely collecting and publishing undigested quantitative data, to the new role of investigating man in society and economy. Coghlan reshaped official data, using economic theory and uncanny intuition, into instruments of economic and social analysis. He devised estimates of national product, income, and expenditure; he analysed the distribution of wealth and income; he examined the demographic structure of Australia; he explored the process of urbanisation; he reconstructed the nature and structure of Australian industry; and he concerned himself with Australian institutions, both economic and social. The resulting annual publications, entitled *The Wealth and Progress of NSW* and *The Seven Colonies of Australasia*, were not mere collections of raw data, but were research reports on the results of his investigation into the nature of Australians in their unique society and economy. And they were reports made accessible to a lay audience. As such they had no peer in their time either in Britain or the USA.

Census taker

Coghlan was no less creative in his role as census taker. The modern system of occupation classification, which draws a distinction between 'breadwinners' and 'dependants', is largely his work. At the Census Conference held in Hobart in March 1890, convened by the various colonies in an attempt to forge a common approach to the forthcoming 1891 population census, the Victorian representatives wanted all colonies to adopt the British method (developed by Dr Farr, Deputy Registrar-General of England), which classified occupations according to a non-economic criterion – the 'material which workers employed in their trades'. Coghlan, who wanted to provide an economic structure for the census, wrote: 'The proposal was strenuously opposed by the author, and, as it fortunately happened by Mr. R. M. Johnston, of Tasmania', because of the anomalies created by the 'British system' – no distinction was made between 'producer and dealer', and the 'domestic' occupation category included 'wives and widows, sons and daughters, relatives, and visitors, university students, children attending school, beer and wine sellers, and persons engaged in the supply of board and lodging, and in rendering personal service for which remuneration is paid'.[69] Coghlan and Johnston proposed the 'Australian method' which began by dividing the population into two groups – 'breadwinners' and 'dependants' – on the basis of whether remuneration was paid. Coghlan explained:

> In dealing with the subject of occupations, as for every other branch of the Census enumeration, it was necessary to account for the whole population, though it is obvious that a large proportion had not what is technically termed an occupation . . . In order, then, to

> account for the whole population, it has been necessary to show it in two divisions, viz., breadwinners, and dependants or non-breadwinners.[70]

He also made a clearer distinction between production and distribution when framing occupations.

The implications of this 'Australian method' are enormous, and are with us still. It is not only the foundation on which modern census occupational categories world-wide are based, it is also the cornerstone of current official national income estimates. Coghlan's distinction between 'breadwinner' and 'dependant' is, in effect, a distinction between market and household employment, which is essential for a wider analysis of total economic activity. Although it has gone unnoticed, this distinction between market and non-market activities influenced the construction of modern national income estimates, and modern approaches to the economy. Coghlan's own estimates of national income are estimates of market income rather than total income, as are the estimates of Colin Clark and Simon Kuznets who followed him and helped to determine the nature of official national accounts throughout the world. Only in the last decade or so has serious consideration been given to estimating non-market income.[71]

Pioneer national accountant, 1886–1903

It is as a pioneer of national accounting that Coghlan is best known.[72] Considerable claims can and have been made for the quality of Coghlan's national income estimates of Australasia. If nothing else, they were the first *official* estimates of national income to be produced anywhere in the world. As a region of recent settlement, Australia experienced rapid development in the second half of the nineteenth century (with real GDP increasing by 5 per cent per annum) and, not surprisingly, attracted the attention of the quantifiers. The first attempts were made by W. C. Wentworth (the famous son of a convict), who estimated the gross value of physical output in NSW and Tasmania for 1819 and 1821,[73] and M. G. Mulhall, who provided crude income estimates in his *Dictionary of Statistics* from 1882.[74] Mulhall's lead was followed by the statisticians of both NSW (T. A. Coghlan)[75] and Victoria (H. H. Hayter).[76] While Hayter's 1891 estimate was a direct copy of Mulhall's, Coghlan's 1887 calculation took a more independent path, which continued to develop in an innovative way until the early years of the twentieth century.

Mulhall may have been the first great populariser of aggregative statistics, including national income, but it was Coghlan who transformed a crude value-added approach into a modern system of national accounts (in gross terms) involving estimates of income received (from 1886), value added (from 1888) and expenditure (from 1888), and who explored the difference between domestic and national product.[77] In addition to improving his methods, Coghlan was able, as Government Statistician, to improve the nature and quality of the raw data. His estimates for 1891 were made more reliable using data from the population census of that year which, as we have seen, was radically restructured through his personal intervention. By the late 1890s, Coghlan had developed his own original system of national accounting using tax data (first collected in NSW in 1897) as well as population, production, and retail data.[78] At this stage he was acknowledged as a world leader in national income accounting, and the quality and sophistication of his estimates were superior to anything published in Australia – by J.T. Sutcliffe (1926), F.C. Benham (1928), Colin Clark and J.G. Crawford (1938) – or elsewhere – Simon Kuznets (1937–40) in the USA and Colin Clark in Britain (1932 and 1937) – prior to the official estimates developed during and after the Second World War.[79] Much intellectual capital could have been saved had Coghlan's achievements been used as a foundation for the revival of interest in national income from the 1930s. As a later member of the Australian tradition, N.G. Butlin, said: 'Coghlan's achievements were remarkable. It is, indeed, extraordinary that his influence did not persist; and that a generation later, economists in Australia congratulated each other on pioneering achievements which Coghlan had surpassed over thirty years before.'[80]

Why was Coghlan's work overlooked? There are many reasons. First, there is Coghlan's decision to refuse the new post of Federal Statistician offered to him in 1906, which, given the lack of interest in economic aggregates of those finally appointed, prevented national accounts from becoming an integral part of Australian official publications until the 1940s. Australia narrowly missed out, thereby, on having a continuous series of *official* national accounts stretching back to the 1880s. But, had he accepted this post, Coghlan probably would not have written *Labour and Industry*, a cornerstone of the Australian tradition in economic history. The gain for economics would have been the loss for economic history. Secondly, Coghlan was a generation or so before his time. Orthodox economics expressed little interest in macroeconomic questions from the work of the classical economists to the *General Theory* of J. M. Keynes. It was the *General Theory* that brought the work of national accountants such as Clark and Kuznets into international prominence. Thirdly, there is the fact that economic ideas, like other ideas, make little international headway until they are 'reinvented' by the superpowers. Finally, this is an outstanding example of the loss of collective memory that is inevitable in a discipline such as economics with little or no interest in the past, even its own past.

A more general question is why in the absence of the theoretical framework provided by the *General Theory*, Coghlan bothered to develop a sophisticated system of national accounts. The answer would appear to be that in order to advise governments about contemporary economic problems, he needed to analyse the rapidly developing Australian economy. To do so Coghlan – like Petty, Davenant, and King before him – had to isolate the main determinants of income, wealth, distribution, and economic 'progress'. The title of one of his official publications, *The Wealth and Progress of NSW*, provides the clue. Coghlan's interests were those of the classical economists, rather than the neoclassical writers, precisely because the interesting issue of the time in 'new' regions was development rather than efficiency. It appears likely that he had in mind a relatively complex growth model derived from Smith, Ricardo, and Malthus. Certainly in *Labour and Industry*, as well as in his earlier official publications, Coghlan was concerned with the relationship between population growth and economic development.[81] The pity is that he adopted an official stance to his work and never commented upon his own methods. Indeed, he is famous for his authoritative and laconic statement in the preface of *Labour and Industry*: 'For the statistics I am my own authority.' No other economic historian has ever been able to make this claim.

Analytical economic historian

This leads us to Coghlan the economic historian, the founder of the Australian tradition. Initial impressions may lead the casual reader of *Labour and Industry* to conclude that Coghlan's *magnum opus* (of 2380 pages) is similar in style to that of J. H. Clapham. While it is true that Coghlan employs a discursive style, it clothes a fundamentally analytical and quantitative approach. There were no other models available at the time anywhere in the world for the analytical study of the historical process of growth. The focus of his history is the wealth and progress of the Australian people, and he examines this important classical issue by employing an implicit growth model. *Labour and Industry* is divided into seven chronological parts, based upon economic rather than political criteria, and within each of these sub-periods he investigates the private market sector – with chapters on immigration, land settlement, labour, wages and prices; the public sector – with chapters on public works and economic policy; and the institutional (particularly financial and legal) framework. It is by using the analytical model of 'progress' that Coghlan is able to provide continuity between sub-periods. This analytically economic structure contrasts sharply with Cunningham's politically determined periods and Clapham's non-analytical pragmatic framework.

The working of Coghlan's analytical model can be illustrated by taking one of his sub-periods, that covering the 'boom and bust' of 1873 to 1893, in which 'The prosperity of the earlier years was coupled with an orgy of speculation which . . . led directly to the great financial crisis of 1893'.[82] Coghlan, who employed his national accounting framework to relate his model to reality, was concerned to analyse the interaction between population, capital, and land:

> From 1871 onwards the colonies had been carried along in a growing tide of prosperity, and wages had been improved without the purchasing power of money being reduced; the import of capital had acted as a wonderful stimulus to the whole community. But the capital absorbing capacity of a country is conditioned by the growth of its population and the room for increased production.[83]

Coghlan used this classical model to examine the process of expansion, or 'progress', in the Australian colonies during the 1880s that led to a downturn after 1889 and financial crisis in 1893. As he argued:

> During the earlier years capital was brought to the colonies in readily absorbable amounts, and this may be said to have continued until about 1885, with excellent results for the whole community. After 1885, in certain years and by certain colonies there was amazing recklessness both on the part of the Governments and of the financial institutions . . . Australia could have absorbed freely and advantageously all the money that it received, if there had been time given for its assimilation, but to bring in seventy-three millions of money to a community numbering between two and three millions, as was done in the five years 1881–85, and 100 millions

> during 1886–90 was to provoke extravagance and invite wastefulness . . . [with the result that] there was more or less wild speculation, an arrest of industry and the withdrawal of large bodies of men from productive pursuits, followed later by a derangement of the labour market, with strikes, lock-outs, and industrial unrest.[84]

Victoria, he claimed, was most at fault – as indeed it is one hundred years later – while 'New South Wales with its large area, more extensive natural resources, and inferior development, was better able to absorb capital'.[85] It was no surprise to Coghlan that the centre of the depression in the 1890s – as in the 1990s – was to be found in Melbourne rather than Sydney. This classical economic focus on capital, labour, and land would appear to be the reason that Coghlan devoted most of his attention to rural development to the neglect of urbanisation, despite the fact that he quantitatively charted urban development, albeit disapprovingly, in his official publications.

Timothy Coghlan was an engineer, mathematician, statistician, demographer, economist, public servant, policy adviser, and diplomat, but it was to economic history that he devoted his last twenty active years. *Labour and Industry* was the culmination of a lifetime of technical training, acquisition of empirical knowledge, and experience of how the economy worked. This was the 'statistical history' that had been promised in March 1888 at the foundation of the journal of the Australian Economic Association, of which Coghlan was a leading member. It was also meant to be the key to understanding man in economy and society – a quest that had led Coghlan from engineering to 'statistical history'. It was an understanding that could not be achieved through abstract thought in economics and mathematics in isolation from empirical reality. Coghlan had no doubt that the social sciences in general and economics in particular must be an historical study.

E. O. G. Shann (1884–1935) – academic pioneer

Edward Shann is the other major pioneering figure of the Australian tradition. While he cannot stand alongside Coghlan as a quantitative and analytical economic historian, Shann was responsible for giving academic shape to the discipline. Shann's training was in history and political economy at the University of Melbourne (1901–04) and in political economy at the London School of Economics (1908–10).[86] While he also read widely in politics, sociology, and philosophy, he had little interest in mathematics and statistics. From 1911 to his premature death in 1935, Shann held academic positions in Australia at the University of Queensland (Lecturer-in-Charge, History and Economics, 1911–12); at the University of Western Australia (Professor of History and Economics, and later Professor of Economics, 1913–34); and at the University of Adelaide (Professor of Economics, 1935). He was also seconded in 1930 by the Bank of NSW (now Westpac) to establish the first private-sector economics department, and he contributed actively to a number of key committees advising governments during the Great Depression, just as Coghlan had done during the depression of the 1890s. Indeed, Shann was one of a small group of economists who had a significant impact upon monetary (bank) and fiscal (government) policy in the first half of the 1930s. Despite this active academic life (he was also Vice-Chancellor of the University of Western Australia, 1921–23) and public career, Shann found time to write a number of influential books, such as the prophetic *The Boom of 1890 and Now* (1927) and the imaginative *An Economic History of Australia* (1930) which became the discipline's major text book (Coghlan's was too long and detailed for this purpose) until the mid-1960s.

The Fabian at the LSE

Just before the First World War even a casual observer of Shann would have anticipated his adherence to the British tradition in economic history. His training at the LSE from 1908 to 1910 under the supervision of Graham Wallas brought him into close contact with the Fabian version of the historical school of economics. There is little doubt that he was strongly influenced by this tradition prior to his return to Australia. His letters from London in 1909, for example, show a keen interest in the living conditions of the working class in England, France, and Germany, and when the Poor Law Commission Report emerged in that year he claimed to be 'wolfing it as fast as I can read'.[87] Shortly thereafter he was employed on a part-time basis to construct an index for the Minority Report of the Poor Law Commission.[88] And when attending a Fabian summer school he wrote: 'The Fabian leadership impresses me very much . . . they are, I think, the most important forward influence in English civilisation'; and he mentions aspiring to 'intelligent citizenship in a socialist state'.[89] He also urged his brother back in Australia to 'Borrow some Fabian tracts . . . and cogitate on their analysis of the relations of the classes', and argued that Australia's future lay with a re-educated Labor Party through 'persuading labourites to be broader minded socialists'.[90] The point that requires emphasis is that Shann at this time adopted Fabian attitudes towards government intervention in the market economy. This comes out most clearly in a very interesting observation he made concerning Alfred Marshall after a visit to see the ageing neoclassicist: 'Marshall is a fiery old man, and a bitterly doctrinaire free trader in spite of his restrained tone in his book. Like many Englishmen he seems to imagine verbal recognition of the superiority of political construction to the economic tendencies of individual self interest quite sufficient.'[91] These are the views of a committed historicist of the Cunningham and LSE variety.

The neoclassical economist

Those who have read Shann's published work will be surprised by this earlier position. It is well-known in Australian circles that, during the inter-war period, Shann was the outspoken advocate of neoclassical economic policies – of freely operating factor and commodity markets – and the scourge of the 'apostles of restriction' who urged, and were responsible for, tariff protection, agricultural subsidies, and wage fixing through the Arbitration Commission. Of the labour market, for example, he wrote: 'Fluctuating wages have a social function to perform in minimizing unemployment and sending labour to Sydney or the bush';[92] and of the commodity market he argued:

> the prices that direct the goings and comings of sound prosperity are the signs by which men learn from willing buyers their changing needs . . . A government or board of control that seeks to fake the world's prices does so at the peril of the citizens and producers whom the faked prices mislead. It deranges and weakens its whole economy if it forces them to work with costs that the outer world is under no obligation to meet.[93]

There is nothing in these words – colourful and passionate as they are – to suggest that Shann had ever been a Fabian or a historicist. Indeed, Shann's memory was recently revived in Australia by the new right, not suspecting that he had once been an interventionist.[94] As we shall see, Shann is not the only Fabian in the Australian tradition to renounce his interventionist views after studying overseas.

Shann's most well-known scholarly work is *An Economic History of Australia* (1930). Like all Shann's published work, it was written during his mature phase of neoclassicism, and hence it is firmly in the Australian analytical tradition. The influence of Coghlan is explicitly acknowledged in the preface, as Shann wishes to 'make abundantly clear his obligations to others, and above all to Sir Timothy Coghlan, the first to labour in these fields'. Not surprisingly, it is to Coghlan that Shann turns when seeking to quantify his history. While his history is much shorter, more readable, and more conventionally 'historical' than Coghlan's, it is shaped and controlled by the implicit use of neoclassical economic theory. What makes it readable is its strong, almost overpowering, theme of the individual's struggle against the forces of restriction, both natural and man-made. This involves a highly imaginative fusion of the central focus of neoclassical economics (economic man), of the problems facing British immigrants in alien and threatening new environments, of the general condition of mankind (as Shann saw it), and of Shann's own unequal struggle with life.[95] I know of no other economic historian who was able to achieve such a multilayered explanation of man in society involving theory, history, philosophy, and personal psychology.[96] Shann may not have been one of the most technically prepared analytical economic historians but he is certainly one of its most imaginative.

Shann's history of Australia is divided into three sub-periods – 1788 to 1860, 1860 to 1900, and 1900 to 1929 – on an economic rather than a political basis. The first of these periods examines the struggle of white settlers (both public and private) with a strange and unpredictable environment[97] – a struggle that, after numerous setbacks, is ultimately successful in developing the major land-based export industry of wool. It is an account in which the forces of progress are the enlightened administrators, explorers and entrepreneurs who, in the pursuit of 'self-interest and hope' were able to overcome the oppression of both an alien environment and of the 'apostles of restriction' which included unimaginative bureaucrats and dangerous interventionists.[98] This was a period of transformation from a primitive form of 'communism' (of the centrally determined kind, as in the Communist USSR) to a small but viable free enterprise system that was able to respond to the gold rushes of the 1850s.

In the second economic phase of boom and bust, 1860 to 1900, Shann employs a theoretical framework similar to Coghlan's, by which he focuses upon the productive resources of land, labour, capital (mainly transport facilities), as well as forces of restriction, such as David Syme and his protectionist friends. When Shann examined the third period of slower expansion that – in January 1930 when the preface was written – was already plunging into the depression he had predicted three years before, the strong theme of individual struggle against restriction had become overwhelming. Accordingly, the chapters in that section are devoted to a passionate attack on 'wage fixing', tariff protection, and 'protection all round'. This polemical focus can be explained by his conviction that the late 1920s were duplicating the conditions of the late 1880s – namely, an excessive dependence on overseas capital at a time of falling export prices – and that a serious depression would emerge in the 1930s. He correctly foresaw the Australian growth process about to break down into its second major depression in forty years.

There is in Shann's history an implicit growth model. Basically, it is a synthesis of the classical and neoclassical systems. Shann was interested in the big classical questions of economic growth, stability, and distribution, but he viewed them from the neoclassical perspective of market efficiency. He thought that the optimal rate of progress could be achieved, and economic crisis avoided, if factor and commodity markets were allowed to operate freely. Only if a nation's institutional structure allowed the entrepreneur full scope to pursue his own 'self-interest and hope', and thereby introduce new ideas into the production process, could uninterrupted growth become a reality. As soon as the state became involved – either directly by displacing private initiative or indirectly by legislating concerning the operation of markets – misallocation of resources would occur, domestic production costs would rise, and the growth process would break down. This had occurred in the 1890s and would occur again in the 1930s. It is reasonable to speculate that had Schumpeter's work been known to him it would have found a sympathetic response.

The Butlin brothers

The Australian tradition was carried forward vigorously by two brothers of the Butlin family – Sydney James Butlin (1910–77) and Noel George Butlin (1921–91).[99] Their research work dominated the field of Australian economic history during the 1950s and 1960s, and their example influenced a younger generation of scholars, who have not restricted themselves to Australian topics, in the 1970s and 1980s. Their work was neatly complementary, with Syd specialising in monetary economic history and Noel focusing upon change in the real sector of the economy. They had a great deal in common in terms of their training and the general methods they employed in economic history. Both graduated with first-class honours degrees and university medals in economics at the University of Sydney – Syd in 1932 and Noel in 1942. After graduating, Syd became a scholarship student at the University of Cambridge from 1932 to 1935, where he obtained an additional honours degree under the tutorship of Dennis Robertson, who was interested in the monetary side of the economy and who was becoming increasingly critical of Keynes. When Syd returned to the University of Sydney he expected to follow a career in monetary economic theory, but he ended up, possibly owing to inadequate preparation in mathematics – a common experience for all Sydney graduates in economics under Professor R. C. Mills, an institutionalist – becoming increasingly interested in monetary history. Noel, who was a founding member of the NSW Fabian Society, followed a different course, possibly owing to the impact of war on his career. After his graduation in 1942, Noel was reluctantly drawn into the public service by the war and post-war reconstruction. He visited Britain and the USA in an official capacity between 1943 and 1945, when he was exposed to the leading economists of the day, including Keynes, Robbins, Robertson, and Stone. As soon as he was able to get away he returned to the University of Sydney as a lecturer in economic history (1946–49), and then went to Harvard as a Rockefeller Fellow (1950–52) to study under Schumpeter. As Schumpeter died within a few months of his arrival, and as the Centre for Entrepreneurial History became increasingly under the influence of sociologists, Noel lost interest and returned to Australia to begin a forty-year association with the Institute of Advanced Studies at the Australian National University. Although the USA at that stage had little to offer a bright young economist schooled in the Coghlan tradition, his experience of the world's leading market economy led to the rejection of his former Fabianism.[100]

S. J. Butlin – monetary theorist turned historian

Sydney Butlin's major work in economic history was *The Foundations of the Australian Monetary System, 1788–1850*, published in 1953. Although he also published a history of the Australia and New Zealand Bank, two volumes (the second with C. B. Schedvin) on the Australian economy during the war, and had, by the time of his death, completed five chapters of a sequel to *Foundations* for the period 1850 to 1914, and seven chapters of the history of the Reserve Bank of Australia (to be completed by C. B. Schedvin), *Foundations* will remain as his scholarly memorial. *Foundations* is a major contribution to analytical and quantitative (in the pre-econometric sense) economic history. Although it is not overtly theoretical, it is a highly organised and carefully argued piece of economic work; and although it does not employ sophisticated statistical techniques (a methodology to which he was opposed), it is firmly based upon the meticulous collection and tabulation of banking time-series data – indeed the statistical appendix occupies almost 20 per cent of the entire volume of 705 pages.

The book's structure is a workable compromise between a chronological (based on economic rather than political criteria) and an institutional (savings banks, foreign exchange market, banks and government policy) approach. While there is no overall model – in fact he warns against 'facile generalisations and "interpretations"' apart from that embodied in the narrative – Sydney felt that his study 'should throw light on the general theory of economic development'.[101] He thought there was something to learn from analysing an economy from its beginnings, because the existing theory of economic development by Marx and Schumpeter 'presupposes a fully developed economy of a capitalist type'. But these lessons were to be left to the perceptive reader, as Syd's work is characterised by intentional understatement. In reviewing the 1969 reprint of Coghlan's *Labour and Industry*, Butlin reveals something of himself in this respect: 'The enduring impression . . . is of the best of human minds assessing and questioning the world around him while never allowing himself to override the constraints of hard measurable reality.'[102] Although Syd Butlin may have been almost pedantically concerned with getting the historical facts right, his sharp focus upon the real world rather than theoretical constructions, despite his thorough theoretical training, is a major characteristic of the Australian tradition. Yet his main regret was 'that the threads of monetary development are imperfectly woven into the fabric of the growth of the Australian economy', owing to the need to go back to the 'original sources' to get the detail right.

As with other pioneers of the Australian tradition, and in contrast to those in the British tradition, Syd Butlin was an economic determinist. As he said in the opening sentence of *Foundations*:

> Australian economic history is the major part of all Australian history; from the beginning economic forces have dominated development in a way that should gladden the heart of any Marx-

> ist. What is true of any particular strand of economic growth – land settlement, labour relations and labour organization, immigration, secondary industry – is also true of each major stage in the development of the community as a whole: each is characterized by economic changes which conditioned political, social and cultural changes.[103]

Sydney Butlin, like others in the Australian tradition, provides not only an economic and quantitative framework for the reconstruction of the past, but also an economic explanation of the progress of human society. In contrast, the British tradition attempts to explain economic 'progress' in non-economic terms. Prior to the emergence of Keynesian and neoclassical growth models, the Butlins looked for inspiration amongst the classical economists and Schumpeter, rather than the English historicists.[104]

N. G. Butlin – analytical historian and economic nationalist

Noel Butlin's work in economic history presents us with a paradox.[105] No one who has read his wide range of publications can fail to be impressed by the power of the man's intellect (despite his difficult prose style), or fail to wonder why he focused that intellect so single-mindedly on one small geographical area – the economy of Australia. It is a focus that has denied him (as it denied Coghlan and Shann) the wider international recognition that has been given to lesser scholars elsewhere. The reasons are undoubtedly complex, but it probably had much to do with an intensely felt sense of national identity – he believed that Australia really does matter in a world context and that it has to a significant extent determined its own destiny. He shared this feeling with a number of very talented Australians of the same, or a slightly earlier, generation – men such as H. C. Coombs, Roland Wilson, Jack Crawford, Trevor Swann, Jack La Nauze, Manning Clark, and, at the end, Keith Hancock – who were determined that after the war they would help to build a new Australia that would take an active and independent role in world affairs. These men could have been successful on the world stage but instead they chose to stay in Australia in academic and public service. This sense of economic nationalism pervades Butlin's work, particularly that which emerged from the 1950s and 1960s.

In the Australian context, however, Noel's interests were broad, ranging through pioneering work on historical national accounts, quantitative and analytical macroeconomic history, the role of government in Australian economic development, the cost of traffic accidents, the impact of urbanisation on the environment, demographic work on the Aboriginal population, the Aboriginal economy during the 60 000 years prior to white settlement and, up to the day he died, an economic history of black and white settlement before 1850, which will be published posthumously. Clearly, it is premature to judge his final book (although I have read most of its chapters in early drafts), but the work for which he will mainly be remembered is his historical national accounts covering the period 1788 to 1939,[106] and his impressive analytical economic history, *Investment in Australian Economic Development*. The international reception of this book, which appeared in 1964, the same year as Conrad and Meyer's *Slavery* and Fogel's *Railroads*, was muted not only by its Australian focus, but more particularly by the unnecessary fuss made about the 'arrival' of cliometrics. Butlin's book was a major and enduring statement made in an analytical and quantitative tradition that stretched back some eighty years, whereas Conrad, Meyer and Fogel were taking the first tentative steps in establishing an analytical and quantitative approach in the USA which, once all the irrelevant hype about counter-factuals (that have always been a part of historical study), hypothesis testing, and primitive econometrics is swept away, is basically similar to the much older Australian tradition. There can be no doubt as to which of these books published in 1964 makes the greatest contribution to our understanding of the process of economic development – Butlin's book, which reflects the longevity of the tradition of which it is a fundamental part, is a far more mature, comprehensive, and complex analysis than those of the American authors.

Noel Butlin's *Investment* draws together the main strands of the Australian tradition initiated by Coghlan and Shann. He takes up in the 1950s where Coghlan left off in the early 1900s, by linking the official Australian estimates of GDP and investment that resumed in the 1940s, with the national income estimates of Coghlan (although they were reworked on a different basis) and beyond. As a result, Australia is probably the only country to possess a complete set of annual estimates of national income for its entire written history. As Butlin, unlike his mentor Coghlan,[107] was not his 'own authority' for the raw statistical data, he was forced to pioneer new methods – using production data for the primary and secondary sectors, and income data for the tertiary sector.[108] This pragmatic approach has resulted in a set of national accounts of relatively high quality owing not only to the quality of official data pioneered by Coghlan, the official statistician, but also to the analytical and quantitative abilities of Butlin.

Like the work of Shann, Butlin's *Investment* contains a dominant and shaping interpretation. Butlin's theme is that Australian economic development in the second half of the nineteenth century was determined largely by Australian rather than by British decision-makers, in the light of Australian economic conditions, and that the deep depression of the 1890s was a logical outcome of this development process. The long boom from 1860 to 1890 (which was a process involving both private and public entrepreneurs laying down urban and rural infrastructure by employing both domestic and foreign capital) sows the seeds of its own destruction as, progressively, local investment criteria deteriorates, speculation emerges, and the old leading sectors starve new sectors of funds. Given Butlin's early interest in Schumpeter's work, it is not surprising that he would tell a story about the internal mechanics of boom and bust – a process of disequilibrium – in which local entrepreneurs, both public and private, play a leading role. There is also much here that he has in common with Shann, with the novel suggestion, however, that governments can be entrepreneurial. While this interest in the direct economic activities of government can be traced back to Coghlan, it also appears to owe much to Butlin's earlier Fabian interests.[109] Despite this obvious Schumpeterian influence, particularly concerning the process of economic growth, there is little reference to innovation in *Investment*. In part this is because Butlin thought that 'the process which appears central to the success of these efforts [to utilise virgin natural resources was] the creation of physical assets',[110] but also because back in Australia at the ANU, the economics group led by Trevor Swan, and of which Butlin was an integral part,[111] was preoccupied initially with growth models of the Harrod–Domar type, in which attention focused upon the role of investment in maintaining equilibrium growth.[112] In the years that followed, Trevor Swan's department briefly became one of the world's leading centres of growth economics, producing what came to be known as the Swan–Solow neoclassical growth model.[113] But it is important to emphasise that these growth models (which are not really growth models at all) focus upon outcomes rather than processes. As Butlin was attempting to come to grips with the *process* of economic growth, it was to Schumpeter, rather than Harrod, that he turned for inspiration.

Noel Butlin's *Investment* is the most impressive analytical economic history of any of the works discussed in this chapter. The book is divided into four main sections: a quantitative analysis of the rate, nature, structure and sources of growth based upon his GDP and investment estimates; an analysis of developments in the private sector, with chapters on the role of investment in each of the rural, industrial, and residential industries; an examination of the public sector, that focuses upon investment in transport and communications; and an analysis of the process of boom and bust, which examines the emergence of longrun structural disequilibrium. Essentially, Butlin's approach is pragmatic – the book's structure is influenced by Keynesian theory, of both the static and dynamic variety, but the detailed argument in each chapter relies heavily upon neoclassical models. While there are strong similarities in structure, method, and interpretation between Butlin's *Investment* and Coghlan's *Labour and Industry*, the main differences arise from the fact that Coghlan wrote his history without the advantages of other examples of analytical economic history or the dynamic models of Schumpeter or Harrod–Domar. As the similarities are more significant than the differences, it may not be too fanciful to suggest that, had Coghlan written his book in the 1960s, it would have looked a lot like Butlin's *Investment*.

Butlin makes little comment about his methods, but what he does say is quite revealing. For example, he quite rightly comments on the lack of other examples, apart from that of Coghlan, of the type of history he was attempting to write:

> Technical economic history of any country, in the sense of economists' appraisal of actual realised processes of growth, is still unfortunately rare. I have found even less guidance in the few essays which exist elsewhere examining the sequence of economic growth in terms both of aggregate behaviour and the performance of major investment components.[114]

The latter concern was a reaction to the highly simplified Keynesian and neoclassical growth models that were under discussion at the ANU and elsewhere in the 1950s, and the need felt by Butlin to move the debate from the rarified theoretical atmosphere to the real world. This book is probably the most useful and durable outcome of those seemingly endless debates and articles about simple growth models. Butlin's view was that useful theoretical models had to be based upon a sound empiricial analysis of reality. And that meant history. He claimed, for example:

> Nor can analytical study be very meaningful without understanding institutional behaviour. This book is intended to be primarily an institutional and historical account of Australian investment and economic development. I hope to follow it with a third essay, a strictly analytical study of growth.[115]

The first was his volume of national income estimates, the third, the 'strictly analytical study', did not eventuate. Butlin appears to have intended to use his experience with historical processes to write a more realistic volume on the theory of growth.

After completing his analytical history, Butlin reflected upon one of the pioneers of the Australian tradition – Timothy Coghlan. He said:

> Returning to *Labour and Industry* after working in detail over the period, one cannot fail to be struck by the greatness of this work, by the grandeur of Coghlan's mind, the keenness of his appreciation, his essential accuracy of both fact and judgement and his obvious fascination with this theme. No book before or since has matched Coghlan's in Australian history and there can be little doubt that had these volumes been given serious academic attention as they warranted, there would, today, be a much greater appreciation of the problems and processes of change in the Australian economy.[116]

Butlin's book had the impact on Australian history and economics that he claimed as Coghlan's due; the pity is that its appearance was not noticed as widely as it deserved throughout the English-speaking world. But time is on his side.

Part II

Economics in Time

5 Economies 'Lost' in Time

From the depth of knowledge past

(John Milton, 1667)

DIGGING UP THE DISTANT PAST

It is sometimes claimed by general historians, possibly with a degree of satisfaction, that quantitative economic history will always be restricted to the modern period owing to the lack of reliable statistical data before the seventeenth century. As one well-known scholar has recently written:[1]

> Unfortunately, economic statistics from the distant past are very, very poor, or they are simply not available at the macro- or whole society level, so that we cannot go in for real measurement. This is no field for the strict cliometrician, who thus misses working on some fascinating periods . . . or doing much comparative work – at least until cliometricians start compiling more of their own statistical series from primary resources.

Certainly, quantitative economic historians have acted as if this were true, by focusing attention upon the period since the Industrial Revolution. There have been a few exceptions, but even these scholars have merely speculated about the distant past rather than making an attempt to reconstruct it empirically.[2] The outcome is a curious professional divide located around the seventeenth century, with general economic historians on one side and analytical economic historians – although not exclusively – on the other. Not only do the methods of each group differ, their views of the nature of human activity also contrast starkly. Once the history of the human race passes across the threshold of the seventeenth century its values are transformed from a communal to an individual basis. Hence the impression is given, quite erroneously, that the history of humanity is discontinuous.

This historical divide is artificial, misleading, dangerous, and unnecessary. It is artificial because it reflects not the nature of human development, but rather the methods and preconceptions of different traditions in historical scholarship. It is misleading because it gives the impression that there has been a recent and major change in the nature of man. It is dangerous because it encourages the view, particularly amongst natural scientists, that the nature of man can be manipulated in the future. And it is unnecessary, as I hope to show, because it is possible to apply quantitative methods to reconstruct the distant past. Reliable statistical data are available throughout the very longrun; it is just that they have been overlooked, or put to one side as being uninteresting or too difficult, by the non-quantitative explorers of these far distant places. By recovering these records we will be able to dig up those economies that are currently 'lost' in time for the analytical economic historian.

Since the beginning of civilisation, precise and accurate records have been kept regarding state and business revenues, and a surprising number of these various records have survived. Perhaps the most remarkable are the economic data recorded in cuneiform writing on clay tablets during the Babylonian and Assyrian Empires between the nineteenth and ninth centuries BC – almost three thousand years ago. Attempts have already been made to measure fluctuations in real wages between the years 1900 and 1600 BC, although like the Brown–Hopkins real wage index from 1264 to 1954, it can tell us little about economic growth or average living standards because of its unrepresentative nature.[3] Of particular interest to this study is that collection of tablets known as Assyrian Doomsday Book, which is part of a royal survey of Harran in north-western Assyria, discovered in the ancient city of Nineveh.[4] In a strikingly similar way to the Norman Domesday Book, compiled some 2000 years later, this ancient data source records, with remarkable clarity and precision, the owner of each farm or village, its administrative status, together with a list of people known by name, houses, animals (including oxen, sheep, and horses), and land (including orchards, vineyards, and forests). As with the Norman survey, the Assyrian entries are arranged according to the owner of property, and within this hierarchical classification are recorded the farms and villages held by each individual. Also, as in Anglo-Norman England in 1086, the Assyrian king granted tax exemptions on land granted to royal supporters. To those familiar with later taxation records – Roman,[5] Norman, and beyond – this sounds very familiar, and leads one to anticipate a similar purpose and motivation.

Was there any essential difference between Assyrian man, Roman man, medieval man, and modern man? To answer this question we need to unearth existing quantitative data on fiscal operations at the level of the state and the productive unit and on the operation of market forces. It would also be interesting to know how the economic achievement of feudalism compared with that of ancient economic systems on the one hand and Third World countries since the Industrial Revolution on the other.[6] Some excellent work has been done, but a greater concentration of effort by analytical economic historians is required if we are to recover these distant economies lost in time. Assyrian Doomsday Book is some way off, but Norman Domesday Book is readily available.

THE FEUDAL ECONOMIC SYSTEM: DOMESDAY ENGLAND

Domesday Book, which offers major insights regarding feudal economic systems, is the most comprehensive and yet the most neglected statistical source in European history. This owes something to the fact that it falls between two groups of specialists, the Anglo-Saxon scholars who see their period as ending in 1066, and later medievalists, who see the thirteenth century (when the far less comprehensive manorial records become available) as their starting point. Domesday Book has attracted a group of specialists, largely non-economic in focus, who are more interested in the document itself than in the reality that gave rise to it. H. C. Darby, for example, has written: 'We can, however, plot the valuations only with the idea of marshalling the information of Domesday Book rather than with the hope of penetrating through that information to the geographical reality of eleventh-century England'.[7] Accordingly, this data source, which is without peer in the history of Europe, if not the world, has not been used in a systematic way to examine longrun change. In this important sense the Domesday economy is lost in time. An unfortunate outcome of this fact is that when scholars, who specialise in the period from the thirteenth century, comment upon the eleventh century in passing they project a later and very different world on to the Domesday economy.

Domesday Book has an important role to play in our understanding of the past because it can be used to reconstruct the feudal economy of eleventh-century England; to provide a foundation for calculating the rate of growth of the English economy and, by implication, of the European economy over the past millennium; to measure the extent of market forces; and to throw light upon human motivation in the distant past. Each of these issues is explored in this and the following two chapters.

Domesday Book was compiled in 1086 by order of William the Conqueror to document precisely what he had taken from Harold Godwinson some twenty years before, and what had been the impact of the great redistribution of estates from English to French hands. Basically, this remarkable survey of virtually all the manors in England south of the River Tees attempted to discover who held what resources, what 'encroachments' had been made on royal property, and what all this was worth as a basis both for a national taxation system (danegeld) inherited from the Anglo-Saxons, and for feudal extraction. The Survey had much to reveal. In the first place, all tenants-in-chief, including free peasants, in 1086 held their land directly of the king in return for a range of dues and services, of which the main one was the provision of knight service. Secondly, economic control had clearly passed into fewer hands, with twenty barons and twelve ecclesiastical tenants-in-chief holding 40 per cent of the land in England in 1086.[8] This concentration of wealth enabled the king's most trusted followers to supply the large number of knights that he required. One estimate suggests that in the vicinity of 4000 mounted knights had to be provided by little more than 180 of the king's barons.[9] This number was supplemented by the knight service expected from an equally small number of bishops and religious houses. Thirdly, the degree of control over English peasants by their lords had increased considerably. While the number of slaves declined dramatically during William's reign, the class of free peasants was everywhere greatly reduced in numbers and, throughout much of the country, may even have disappeared entirely.[10] The precise nature of this more tightly controlled form of feudalism and manorialism, which emerged between the Conquest and the Survey, was apparently of great interest to the king, as its imprint is boldly stamped upon the structure of Domesday Book.

Organising a feudal survey

To undertake the Survey that was known as the *Inquisitio*, the counties of England were grouped into about seven circuits, which were visited by teams of commissioners who had no material interests in their circuit. These circuits are thought to have been:[11]

1. Kent, Sussex, Surrey, Hampshire, Berkshire;
2. Wiltshire, Dorset, Somerset, Devonshire, Cornwall;
3. Middlesex, Hertfordshire, Buckinghamshire, Cambridgeshire, Bedfordshire;
4. Oxfordshire, Northamptonshire, Leicestershire, Warwickshire;
5. Gloucestershire, Worcestershire, Herefordshire, Staffordshire, Shropshire, Cheshire;
6. Huntingdonshire, Derbyshire, Nottinghamshire, Rutland, Yorkshire, Lincoln; and
7. Essex, Norfolk, Suffolk.

The survey results for circuits 1 to 6 were recorded in Great Domesday Book, and those for circuit 7 were recorded, in greater detail, in Little Domesday Book.

The procedures adopted within each circuit are considered to be reasonably similar. A list of questions was supplied to the tenants-in-chief, who were identified by the hidage (or tax assessment) lists, in each county; the returns from the tenants-in-chief were reviewed in the county court by juries in each hundred consisting of four Englishmen and four Frenchmen; the data verified in this way were recorded by tenants-in-chief for each county in the circuit; and the circuit returns were sent to Winchester, where they were summarised, edited, and compiled into Domesday Book's first volume. The tenants-in-chief of each county are thought to have been supplied with a set of questions similar to those contained in the text of *Inquisitio Eliensis*, which records the survey of the Abbey of Ely's estates in six eastern counties. The questions are:[12]

> What is the manor called? Who held it in the time of King Edward? Who holds it now? How many hides? How many ploughs on the demesne? How many men? How many villeins? How many cottars? How many slaves? How many freemen? How many socmen? How much wood? How much meadow? How much pasture? How many mills? How many fish ponds? How much has been added or taken away? How much, taken together, was it worth and how much now? How much each freeman or socman had or has? All this at three dates, to wit, in the time of King Edward and when King William gave it and as it is now. And if it is possible for more to be had than is had.

As can be seen from the sample entry in Figure 5.1, which is a manor at Codford in Wiltshire, the recorded answers in Domesday Book appear to be responses to a remarkably similar set of questions.

Tenants-in-chief were circulated with the list of questions from the circuit commissioners, and were called upon to present their responses in the county court. Evidence in the *Inquisitio* suggests that these hearings were dealt with by an enlarged county court which, in addition to the circuit commissioners, consisted of the sheriff, the king's barons with fiefs in the county together with their French tenants, the court of each hundred with its eight-man jury, and eight representatives – a priest, a reeve, and six villeins – from each village.[13] Clearly, this was a very big gathering, involving large numbers of people in each county throughout the land. Little is known of the identity of the main participants in these proceedings. It is thought, however, that the commissioners included bishops, lawyers, and the king's barons, none of whom had any material interest in the circuit for which they were responsible. More is known about the juries, as some are recorded in the *Inquisitio*. As previously mentioned, they consisted of eight people, of whom half were Englishmen and half were Frenchmen, and they appear to have come from the rural middle class.

The involvement of so many people – possibly 10 000 throughout England – in the collection, verification, and compilation of the Domesday data suggests a number of conclusions. First, and most obviously, this was a very expensive survey in terms of both the official expenditures and the private opportunity costs of the many who participated, from powerful baron to poor peasant. It is unlikely that such costs would have been undertaken unless the king were convinced that it would result in a higher future income. Secondly, it is clear that the way in which the data were collected made it a very public event involving many people who had intimate local knowledge of all the manors under review. In these circumstances there would have been few occasions when falsification could have passed undetected. This evidence was also given under oath, and the penalties for perjury were probably severe, because falsification meant depriving the king of his rightful revenue. So concerned was William to ensure the accuracy of this record that he even sent out additional agents to ensure that the commissioners were carrying out his instructions carefully and impartially. This is not to say that mistakes did not creep into Domesday Book, either deliberately or by accident, but the method of collection was such that they would have been minimised. The procedures employed in the collection and verification of this unique set of microeconomic data were so exacting that the reliability of the resulting income statistics is higher than that of modern survey data undertaken by either private or public bodies. This can be explained largely in terms of the much greater power that the Conqueror had over the destinies of individuals both in this life and the next (through excommunication), and of his ability to exploit the close and constant surveillance typical of pre-modern societies. Accordingly, the statistical goodness-of-fit achieved by the use of standard economic functions is much greater for the Domesday economy than for the modern economy.

The English economy in 1086

The manor, which was the central institution in the Domesday economy, produced a range of agricultural products both for internal consumption and for exchange with luxury goods and military supplies. Trade between manors within England, and between England and Europe, was facilitated by a network of markets ranging from local fairs to boroughs and large city-ports (such as London and Southampton). The major shaping influence in this economy was a relatively strong central administration provided by the monarchy which, if not the ultimate owner of land (the main source of income and wealth), certainly exacted a number of important services and taxes from landholders.

The manorial economy

The manorial system, although not uniform throughout England, was most fully established in English England – as opposed to Scandinavian England – and particularly in the heart of the ancient kingdom of Wessex.[14] Arable farming was the central economic activity throughout the country, and was based upon the cultivation of crops such as wheat, oats, barley, and rye, which provided both the basic food requirements of the manor and, sometimes, a surplus for the purchase of those luxury goods and military supplies that could not be produced locally. Supplementing this major activity was the raising of livestock. While manorial land was devoted largely to arable activities, other forms of land use were essential to maintain arable production (meadows were a vital source of hay to maintain the plough beasts in winter) and to supplement it (pasture and woodland were essential to the support of livestock). The arable land of the manor was

XXXVII. TERRA WALERAN VENATORIS.

WALERANN ten de rege *COTEFORD*. Erlebald tenuit
T.R.E. 7 geldb p. VI. hid. Tra. e. VI. car. De ea st in dnio
III. hide. 7 ibi. II. car. 7 III. serui. 7 VII. uitti 7 VI. bord cu. III.
car. Ibi molin redd. x. sol. 7 x. ac pti. Pastura dim leu
lg. 7 v. qz lat. Valuit x. lib. Modo. XII. lib.
Walter ten de. W. *ANESTIGE*. Aluric 7 Vluuard tenb
T.R.E. 7 geldb p. VII. hid. Tra. e. IIII. car. De ea st in dnio
.v. hidæ. 7 una v træ. 7 ibi. II. car. 7 II. serui. 7 VI. uitti 7 IIII.
bord cu. II. car. Ibi molin redd. v. sol. 7 XVI. ac pti. 7 XV. ac
siluæ. Pastura dim leu lg. 7 III. qz lat. Valuit 7 ual. c. sol.
Azelin ten de. W. *BVTREMARE*. Octo taini tenuer T.R.E.
7 geldb p. I. hida 7 una v træ. Tra. e. II. car. Valet. xx. solid.
Ipse. W. ten. I. hida in *STANINGES*. Colo tenuit T.R.E. Tra. e
dimid car. Ibi st. VI. ac pti. Valuit. v. sol. Modo. x. sol.
Ricard ten de. W. *CHENETE*. Leueclai tenuit T.R.E. 7
7 geldb p una hida 7 dim. 7 una v træ. Tra. e. I. car. q ibi. e
cu. I. seruo. 7 II. bord. 7 una ac pti. 7 IIII. ac pasturæ. Valuit 7 ual
Azelin ten de. W. *STANLEGE*. Seleuuin tenuit xx. sol.
T.R.E. 7 geldb p una hida 7 III. v træ. Tra. e. I. car.
Ibi st. III. uitti 7 III. bord. 7 x. ac pti. Valuit. xv. sol. m xxx. sol.
Ipse Waler ten *LANGEFORD*. Osulf tenuit T.R.E. 7 geldb
p x. hid. Tra. e. v. car. De ea st in dnio. v. hidæ. 7 ibi. II.
car. 7 v. serui. 7 VIII. uitti 7 IIII. bord. cu. III. car. Ibi molin
redd. xv. sol. 7 xxx. ac pti. Pastura dim leu lg. 7 II. qz lat.
Valuit 7 ual. x. lib.

Figure 5.1 Sample Domesday Book entry: the manor of Codford, Wiltshire

Source: The Phillimore edition of *Wiltshire Domesday Book*.

LAND OF WALERAN HUNTER

1 Waleran holds CODFORD from the King. Erlebald held it before 1066; it paid tax for 6 hides. Land for 6 ploughs, of which 3 hides are in lordship; 2 ploughs there; 3 slaves;
7 villagers and 6 smallholders with 3 ploughs.
A mill which pays 10s; meadow, 10 acres; pasture ½ league long and 5 furlongs wide.
The value was £10; now £12.

2 Walter holds ANSTY from Waleran. Aelfric and Wulfward held it before 1066; it paid tax for 7 hides. Land for 4 ploughs, of which 5 hides and 1 virgate of land are in lordship; 2 ploughs there; 2 slaves;
6 villagers and 4 smallholders with 2 ploughs.
A mill which pays 5s; meadow, 16 acres; woodland, 15 acres; pasture ½ league long and 3 furlongs wide.
The value was and is 100s.

3 Azelin holds BUTTERMERE from Waleran. Eight thanes held it before 1066; it paid tax for 1 hide and 1 virgate of land.
Land for 2 ploughs.
Value 20s.

4 Waleran holds 1 hide in STANDLYNCH himself. Cola held it before 1066. Land for ½ plough.
Meadow, 6 acres.
The value was 5s; now 10s.

5 Richard holds KENNETT from Waleran. Leofday held it before 1066; it paid tax for 1½ hides and 1 virgate of land.
Land for 1 plough, which is there, with 1 slave;
2 smallholders.
Meadow, 1 acre; pasture, 4 acres.
The value was and is 20s.

6 Azelin holds STANLEY from Waleran. Selwin held it before 1066; it paid tax for 1 hide and 3 virgates of land. Land for 1 plough.
3 villagers and 3 smallholders.
Meadow, 10 acres.
The value was 15s; now 30s.

7 Waleran holds LANGFORD himself. Oswulf held it before 1066; it paid tax for 10 hides. Land for 5 ploughs, of which 5 hides are in lordship; 2 ploughs there; 5 slaves;
8 villagers and 4 smallholders with 3 ploughs.
A mill which pays 15s; meadow, 30 acres; pasture ½ league long and 2 furlongs wide.
The value was and is £10.

organised for production in large open fields.[15] While there was some regional variation, it is generally recognised that open fields predominated over a large part of England south of Yorkshire. Throughout much of this area, settlements took the form of nucleated villages, which in turn were the focus of one or more manors.

Manorial production was organised typically into demesne and peasant tenancies, with the various classes of dependent (or bonded or unfree) peasants – villeins, bordars, cottars and so on – receiving small plots of land in exchange for labour services and customary dues. There were also freemen and sokemen who held their land directly from the king, but who were generally closely associated with a particular manor, sometimes paying customary dues. While the dependent peasants worked individually on their own plots, as well as on those of the demesne, the general organisation of production of the open fields was undertaken corporately. Yet it is claimed that there was scope in this corporate system for the exercise of individual initiative and enterprise.[16]

Within this organisational structure, the techniques of agricultural production were fairly similar throughout much of England, with certain exceptions in the fringe areas of the west and north of Cornwall, in the highlands of the Pennines, and the uplands of the highland spine of England, where the focus was on pastoral rather than arable activities. Naturally, within this general technical framework there was local variation in some of the details, such as the balance between arable and pasture, the crop mix, and certain 'rules and institutions' of agriculture.[17] The large open fields were divided into long narrow strips, the shape of which reflected the logistics of employing large plough teams of up to eight oxen pulling a heavy Anglo-Saxon plough. As is well known, the strips of arable land owned by individuals, including the lord, were scattered throughout the large open fields rather than being consolidated into compact holdings. The fertility of these strips, a problem of critical importance in the eleventh century, was maintained by employing a two-field system of rotation whereby one was left fallow each year, and by the addition of animal manure, lime, marl, peat, and compost to the soil.[18]

The urban economy

Although the Anglo-Norman economy was predominantly rural, it supported a number of important boroughs or towns. London, which was omitted from Domesday Book, is thought to have had a population in excess of 10 000, possibly 12 000, at the time of the Survey. Of the other main boroughs, York had in the vicinity of 8000 inhabitants; Winchester (not recorded in Domesday Book), Norwich, and Lincoln had in excess of 6000; Oxford and Thetford had about 5000; Canterbury, Colchester, Dunwich, Exeter, Gloucester, Leicester, Lewes, Nottingham, Sandwich, Stanford, and Wallingford had more than 2000; and Bath, Cambridge, Chester, Chichester, Dover, Hastings, Huntingdon, Hythe, Northampton, Shaftesbury, Ipswich, Maldon, and Southampton had over 1000.[19] In addition, there were a further eighty smaller boroughs recorded in Domesday Book. By 1086, therefore, towns were clearly an integral part of the English economy. Yet their role should not be overstated, as they employed only about 8 per cent of the total population. Urbanisation in England had not proceeded very far before the Survey.

It is difficult to be very precise about the nature of Anglo-Norman boroughs, because the Domesday Survey appears to have been little concerned with them except in so far as they had some relationship with manors. In particular, no systematic attempt was made to record the industrial and commercial resources of even the major towns. Indeed, the Domesday commissioners appear to have been more concerned to record the agricultural rather than the industrial or commercial characteristics of boroughs. For some boroughs there are records of ploughs, arable land, villeins, and bordars, but not of artisans or their workshops.[20] Towns were viewed, therefore, largely within a manorial rather than an industrial or commercial framework. Certainly there are numerous references in Domesday Book to burgesses and town property belonging to the wealthier lords of nearby manors. Manors even had men working for them in far-distant salt-making boroughs, and claimed rights to salt produced in those centres.[21]

A NEW BEGINNING: DOMESDAY MACROECONOMIC DATA

The Domesday Survey is unique. It provides, as we have seen, an almost complete set of microeconomic data for England south of the River Tees, concerning property 'ownership', fiscal responsibility, income, population, land, and capital stock, all in considerable detail. As there are in the vicinity of 30 000 individual observations containing, on average, ten to twelve variables at three points in time – 1066, 1086, and a variable intermediate year – Domesday Book contains up to one million separate pieces of systematic information about the English economy of the eleventh century. I have already used this microeconomic data to examine the nature of the manorial system of production in *Domesday Economy*,[22] and here employ it to estimate national income and per capita income in 1086 and to thereby reconstruct the nature of the English feudal economy. This provides the reliable and comprehensive macroeconomic data that our earlier commentator claimed was absent for economies in the distant past. The wealth of detail in Domesday Book is not only unequalled until the twentieth century, but stands in marked contrast to the handful of fragmented and unrepresentative statistical data that are currently available for England during the later Middle Ages. Ironically, these inadequate data are employed by those more familiar with the period from 1200 to 1500 as a test of the veracity of the comprehensive Domesday Book data. This is the cost of periodisation that N. S. B. Gras warned the profession about back in the 1920s. If there is to be a benchmark in the Middle Ages, that benchmark should be Domesday Book.

The population data

Domesday population and workforce data play an important role in the subsequent national income and per capita income estimates. Abundant detail about the rural workforce in England in 1086 is provided in Domesday Book: as shown in Table 5.1, the workforce consisted of free peasants (freemen and sokemen), bonded peasants (villeins, bordars, cottars, and coscets) and slaves. The free peasantry (about 14 per cent of the workforce) clung precariously to their small independent holdings, mainly in the eastern counties, and to the claims that some of them had to the king's – rather than the manorial lord's – jurisdiction. The largest group were the bonded peasants who worked on the manorial demesne in exchange, not for wages or commodities, but for the land and plough beasts they required to maintain themselves at a subsistence level. Of these it has been suggested that villeins (41 per cent) worked on the demesne for *approximately* two to three days a week in return for fifteen to thirty acres of land, while bordars (30 per cent) worked one day a week for the lord in return for five to fifteen acres.[23] Naturally, there would have been considerable variation in these labour/land exchanges from region to region. The role of slaves (11 per cent), which appears to have declined from 1066, was also important. A more precise picture of the role of different

Table 5.1 Recorded workforce in England, 1086

		000s		%
Free peasants		36.9		13.7
freemen	13.6		5.1	
sokemen	23.3		8.7	
Bonded peasants		197.9		73.6
villeins	109.2		40.6	
bordars	81.8		30.4	
cottars	5.2		1.9	
coscets	1.7		0.6	
Slaves		28.2		10.5
Others		5.8		2.2
Total		268.8		100.0

Source: Derived from Darby, *Domesday England*, p. 63.

types of labour, together with their contribution to the manorial economy, can be obtained from the application of production functions to the Domesday data as in *Domesday Economy*.[24]

In order to convert these rural workforce figures into population totals, it is necessary to make a number of adjustments, taking into account suggestions made by H. C. Darby.[25] First, we need to include the number of tenants-in-chief (1100) and the number of under-tenants (6000), both categories being separately documented in Domesday Book. Secondly, it is necessary to multiply the total number of heads of households by the average size of households (which were not necessarily families) in 1086. Most Domesday scholars consider that the average household consisted of 4.5 to 5.0 people, and evidence for later periods suggests that a household size of 5.0 might be appropriate. To test the sensitivity of the final per capita income results to variation in household size, three alternatives – 4.5, 5.0, and 5.5 – have been employed in Table 5.2. Thirdly, the incomplete record for Lancashire, and the absence of any record for Cumberland, Westmorland, Northumberland, and Durham must be allowed for. Darby has estimated the rural workforce for Lancashire to be 1800 and that for the four northern counties to be about 5000. Finally, the total size of the urban population has been estimated by Darby to be 120 000.[26] Table 5.2 provides three alternative population totals for England in 1086, ranging from 1.390 to 1.672 million (a difference of 20.3 per cent). The available evidence concerning household size,[27] suggests

Table 5.2 Alternative population estimates for England, 1086

Categories	(1) 4.5	Household size (2) 5.0	(3) 5.5
1. Recorded peasants	1 207 256	1 341 395	1 475 535
2. Tenants-in-chief (1 100)	4 950	5 500	6 050
3. Under-tenants (6 000)	27 000	30 000	33 000
4. Domesday Book total	1 239 206	1 376 895	1 514 585
5. Lancashire (1 800)	8 100	9 000	9 900
6. Northern counties (5 000)	22 500	25 000	27 500
7. Urban population	120 000	120 000	120 000
8. Total for England	1 389 806	1 530 895	1 671 985

Source: Derived from Darby, *Domesday England*, p. 89: Category 6 has been allowed to vary; and his guess (5%) for omitted peasants has been excluded (see Chapter 5, Note 25).

that the data in column (2) – a total of around 1.531 million – is the most plausible, and this has been used throughout the paper. The implications of using the other alternatives are discussed in the Appendix. Darby raises an added complication: were slaves included as heads of households or as individuals? The most likely answer, as F.W. Maitland and common sense suggests, is that slaves were recorded in the same way as peasants. Certainly, there is nothing in Domesday Book to suggest otherwise.

The income data

The overriding objectives of the manorial economy of Domesday England were to produce a surplus that could be employed: to provide a fully equipped feudal army to secure the country against invasion; to satisfy a warrior king's desire for further wealth and power through conquest; and to meet the consumption and investment demands of the ruling elite. Not surprisingly, the surplus, which was known as the *valet*, was given pride of place in Domesday Book. This is not to say, as some have done, that such a system prevented economic growth because of a lack of entrepreneurial spirit by the ruling elite.[28] Only part of the surplus was used for military purposes, and even this provided economic benefits such as the security that was essential to enable systematic investment in fixed assets, and also the possibility of an increase in wealth and income through conquest. This was not just an 'elementary governmental function' as some have argued;[29] it was fundamental to the economic viability of Norman England. Also, the need to maintain a feudal army based upon the mounted knight encouraged investment in meadow-land development (with its vast network of irrigation channels) to maintain the warhorses. This meadow land was also essential for building up the stock of plough beasts, the main item of rural capital. There is even some evidence to suggest that tax concessions were provided on meadow-land development.[30] Further, as an efficient economic system was required to meet military expenditures – the knights were granted manors for this purpose – there was an incentive to invest part of the remaining surplus in the best agricultural practices. The English feudal system, therefore, was conducive to economic growth, with the ruling elite playing an important entrepreneurial role.

In late-eleventh-century England, all feudal services were paid for, not with income or goods but with the means to produce income – rural holdings complete with capital and dependent peasant labour. This system began at the top with an exchange of estates for clerical, administrative, judicial, and military services between the king and his tenants-in-chief, and permeated down through the feudal 'pyramid' to individual manorial lords. In other words, clerics, military officers, administrators, county and central court officials, and other agents of the king were all paid in the form of property rights over rural manors. It was not until the twelfth century that bastard feudalism – in which the larger tenants-in-chief employed growing retinues of retainers – began to emerge and then to accelerate, particularly during the fifteenth century.[31] By 1688, the range of paid occupations was widespread and therefore had to be included separately in Gregory King's accounts. Even within individual manors, the unfree peasants were compelled to provide labour services of various types – including agricultural, craft, and transport – in exchange for the land and tools required for their subsistence. Hence the services produced in English feudal society, outside the small number of scattered towns that must be dealt with separately, can be captured in an appropriate measure of the net return from working the land.

The manorial system consisted of two parts – the demesne economy which provided the manorial lord and free peasant with a surplus, and the subsistence economy which sustained the *unfree* labour that worked on the demesne. It is essential to realise that the income of free peasants, who lived somewhat above subsistence level in all but the worst years, is recorded in Domesday Book in the annual values, and hence is included as part of the demesne economy. Only unfree peasants are included in the subsistence sector. This was a dependent form of agriculture in which bonded or unfree peasants worked on the lord's demesne in return for land and capital that were just sufficient to maintain them and their families at the minimum consumption level *that would ensure the maximisation of the manorial surplus* and a steady growth of population. In 1086 this approached physiological subsistence, and even for the fourteenth century scholars talk of the 'knife edge of subsistence', and the 'peasant art of starvation'.[32] The lord was able to enforce this one-sided contract on his dependent peasant workforce, despite the fact that labour rather than land was the scarce factor of production, because of his military monopoly and the monopsony arrangement (where there are few buyers) with his fellow barons and knights. In the process the manorial lord was able to maximise the economic rent he could extract from labour. The economic position of the bonded peasant, as far as the lord was concerned, was little different from that of the plough beasts – the surpluses over and above the costs of maintaining labour and capital (in both cases food and shelter) were extracted through the manorial system.

This form of economic management is consistent with my argument in Chapter 6 that manorial lords were economically rational.[33] The manorial system was a form of dependent agriculture precisely because labour was the scarce factor of production, and the only alternative to the threat of force, both military and ecclesiastical, by manorial lords was competition between them. The latter, which would have increased the return to labour and reduced the manorial surplus, would not have been a rational undertaking by manorial lords within the economic circumstances of the late eleventh century. Also the threat of force was an effective method of surplus extraction, because its use was largely without cost to manorial lords in a society geared up for war, because the manorial system was capable of close supervision, thereby minimising shirking, and because the need to survive provided peasants with an incentive to complete their work on the demesne as quickly and as effectively as possible so as to be released to work on their own land.

The surplus generated by the manorial system is recorded in Domesday Book as the 'annual values' (*valets*). This surplus can be thought of as gross manorial production on all land encompassed by the manor minus intermediate goods (such as grain seed), and those goods produced on the manor (such as fodder and rough shelter for livestock; food, clothing, and shelter for slaves; and the subsistence output of the peasants' land) in order to maintain manorial resources. This is, of course, equivalent to the microeconomic definition of value added. To put it another way, the annual value is the return to the manorial lord from owning or controlling the resources of the manor, and it includes a return for risk, together with economic rents and quasi-rents extracted from factors that were scarce or in fixed supply in the shortrun.

To translate this microeconomic concept of value added into modern national income terms the usual convention has been followed of adding back into 'value added' the cost of maintaining human labour. Accordingly, the manorial component of national income in 1086 must include not only the demesne income (or the annual values) but also unfree peasant subsistence income (see Table 5.8). On reflection it becomes obvious that this convention is a product of modern humanist thinking: modern democratic society is concerned with the aspirations and living standards of all human individuals. In contrast, the objective function of feudal society did not include the aspirations of non-free labour, either slaves or bonded peasants. In these terms the different perspectives of modern society and feudal society can be measured by the value of unfree peasant subsistence income presented in Table 5.8 on page 194.

Manorial income – the demesne economy

The total value of demesne income, which includes that for all manorial lords and free peasants, can be obtained from Domesday Book by aggregating the annual values for some 30 000 rural holdings. The underlying data and method are far more reliable and robust than those used by Gregory King – basically estimates of numbers of families and estimates of average family income – some six hundred years later. But there are marginal adjustments that must be made to the aggregated total of annual values (or domestic income) for rural holdings in Domesday Book to cover a number of relatively small omissions. These omissions include the four most northern English counties, and the urban centres which included about 8 per cent of the population in 1086.

As the detailed task of aggregating the annual values, or demesne income, requires the exercise of value judgements owing to the problem of allocating individual manors within estates to particular counties, the totals obtained by different scholars will always be *slightly* different. In Table 5.3, for example, the totals obtained by Maitland and Darby differ by £264, or 0.37 per cent. Such minor discrepancies, which make no significant difference to estimates of per capita income, can safely be ignored. In this study, Darby's carefully aggregated total for demesne income – £71 573 – has been adopted.

Manorial income – the dependent peasant subsistence economy

It can be inferred from later evidence that the subsistence sector (which includes only unfree peasants) involved a mix of agricultural and non-agricultural activities, which were reflected in the variation of individual economic 'responsibilities' (days of work on the lord's demesne) and 'rights' (land use). This in turn suggests that we can think of the subsistence sector as a 'closed' economy (see Figure 5.4 on page 203) in which there is a degree of economic specialisation and exchange, with the only constraint being that the total amount of land and capital allocated by the manorial lord is sufficient to sustain the bonded peasant population. Domesday Book, however, does not allow a direct calculation of the value or structure of subsistence output, precisely because it did not provide a taxable surplus. But the Conqueror's great survey does provide sufficient information to enable a calculation of the net output required to maintain the unfree labour force – the villeins, bordars, cottars and so on, and slaves – at a level which enabled manorial lords to maximise their incomes. And it contains income data for the holdings of free peasants that can be used to check my estimate of per capita subsistence income for the unfree peasants. As already discussed, the consumption level of unfree peasants is thought to have approached physiological subsistence in the late eleventh century. What is required, therefore, is an estimate of the value of the subsistence requirements of bonded peasants and their families in terms of food, clothing, and shelter. This estimate is a viable alternative – but only before the twelfth century, after which retained surpluses for formerly bonded peasants began to emerge – to the usual method of calculating the total net output of land (of which we have no record) allocated by manorial lords to their bonded peasants.

The major component of subsistence income is food. An estimate of the value of food consumption can be undertaken by (i) reconstructing the structure of the average Domesday peasant household; (ii) estimating the expected energy requirements of that household; and

Table 5.3 Manorial value added in England by county, 1086 (£)

County	*Maitland*	*Darby*
Kent	5140	4770
Sussex	3255	3116
Surrey	1524	1533
Hampshire	3415[d]	3415
Berkshire	2384	2524
Wiltshire	4770[d]	4770
Dorset	3360[e]	3110
Somerset	4161[e]	4361
Devonshire	3221	3145
Cornwall	662	670
Middlesex	754	740
Hertfordshire	1542	1458
Buckinghamshire	1813	1947
Oxfordshire	3242	2878
Gloucestershire	2827	3204
Worcestershire	991	969
Herefordshire	1115[d]	1115
Cambridgeshire	1847[d]	1847
Huntingdonshire	865	827
Bedfordshire	1097	1164
Northamptonshire	1843	1744
Leicestershire	736	842
Warwickshire	1360	1409
Staffordshire	517[e]	449
Shropshire	852[d]	852
Chestershire	247[d]	247
Derbyshire	461	430
Nottinghamshire	731[d]	731
Rutland	–	–
Yorkshire	1084[d]	1084
Lincolnshire	3253[d]	3253
Essex	4785	5047
Norfolk	4155	4094
Suffolk	3828[d]	3828
All counties	71837	71573

Notes: 1. The figures reported by Maitland are based upon Pearson, and Eyton (e). Gaps have been filled from Darby (d).

2. Differences in county totals reflect the fact that estates were composed of widely distributed manors, which are sometimes difficult to locate.

Sources: Maitland, *Domesday Book and Beyond*, pp. 401–2; Darby, *Domesday England*, p. 359; Pearson, *History of England during the Early and Middle Ages*; and Eyton, *Key to Domesday: An Analysis and Digest of the Dorset Survey*.

Table 5.4 Hypothetical structure of the subsistence household (number of persons)

	(1)	*(2)*	*(3)*
Head	1.0	1.0	1.0
Wife	0.7	0.7	0.7
Children	2.42	2.80	3.17
Older adult	0.38	0.50	0.63
Total	4.50	5.00	5.50

Source: See text.

(iii) calculating the cost of meeting those requirements. The household structure in Domesday England is reconstructed in Table 5.4 using three different assumptions (in order to test the sensitivity of the results) concerning household size: of 4.5, 5.0, and 5.5 individuals. Justification for these assumptions is given in the section on population. The identification of types of individual within these households is based on common sense and detailed English evidence for the late thirteenth century[34] and the period 1564 to 1851.[35] Households could not be maintained for long without a male adult head, but it was common for households to include children but no wife, largely because of the dangers of childbirth in medieval society. Of the remaining household members there would have been survivors from the previous generation (they cannot have been a particularly large proportion, however, because even in the mid-thirteenth century the life expectancy of the 'more substantial peasants' at the age of 20 years was only a further 20 to 24 years),[36] other relations, boarders, and children.

The energy requirements of the Domesday household can be estimated with reasonable confidence by consulting the large literature on twentieth-century experience in developed and Third World countries, and making allowances for the improvements that have occurred even in the latter regions over the last 900 years.[37] Data from developing countries show that the contemporary average calorie intake is 1970 for South Asia, and not much above 2000 for Southeast Asia, China, and Latin America, while individual countries within these regions experience per capita calories intakes of less than 2000.[38] By narrowing our focus to the village level in Latin America, we find that 'pure' subsistence villages have mean nutrient intakes of less than

1700 calories per day.[39] Despite these impoverished conditions, standards in those countries today are significantly higher than in the past. But how much higher?

The only substantive evidence on consumption standards in the medieval records concerns the maintenance agreements entered into by new tenants with retiring tenants. C. Dyer has examined 141 such agreements for the period 1240 to 1458.[40] Although the median amount of grain negotiated in those agreements is 1½ quarters per person per annum (equivalent to about 1710 calories per person per day), the range of observations was very wide – from 0.13 to 3.13 or more quarters. Also, it is clear from some of these agreements that the grain provided in this way contained a considerable surplus for exchange rather than purely for consumption. Nevertheless, this data is broadly consistent with my estimates of subsistence household consumption in Table 5.6 (on page 190). In an entirely different context, N.G. Butlin has attempted to estimate the energy requirements of Australian Aboriginal families (a mix of infants, children, adults, and the elderly) prior to European contact, and claims that 1200 to 1500 calories is an appropriate daily range.[41] While there may be some differences between the minimum energy requirements of 'pure' hunter-gatherers and early medieval agricultural workers, these differences will not be very great. In agricultural societies the peak demands made by ploughing and harvesting activities may have been greater (perhaps in the vicinity of 2500 calories in the eleventh century for adult males) than the energy requirements of hunting and gathering in Aboriginal economy, but these demands were highly seasonal and probably averaged out to comparable levels over the course of a year. Indeed, food consumption in medieval society went through a very marked 'annual cycle of temporary indulgence and real deprivation, with a feasting season from autumn until Christmas, and belt-tightening in the spring and early summer induced more urgently by the reality of short supplies than by the religious obligations of Lent'.[42] Also the differences in energy requirements during winter between these two types of society would not have been too great owing to the very cold winter temperatures (frequently below zero) experienced throughout much of eastern, south-western, and southern Australia, and to the lack of storage facilities in nomadic economies requiring hunting effort throughout the winter months.

Finally, we need to consider the impact in the eleventh century of famine (both natural and man-made) and disease (both human and animal) on calorie intake, which had the effect of driving the average down to and below bare subsistence. The Conquest (1066), the Conqueror's reprisals (1070), and the scorched-earth policy imposed in coastal regions following rumours of invasion had a major impact upon consumption levels. During these periods, as in later centuries, the peasants were reduced to eating acorns and roots.[43] As C. Cipolla has said, in Europe before the Industrial Revolution 'the mass lived in a state of undernourishment'.[44] And, more recently, R. Fogel refers to 'chronic malnutrition' in the pre-industrial period, and compares the nutrition of English labourers as late as the early nineteenth century with the 'lower classes' in India, Pakistan, and Bangladesh shortly after the Second World War.[45] In the context of famines, it is useful to realise that modern relief programmes, which aim to sustain life during the short term, are based upon the average consumption of a pound of cereals per day, which works out at about 1140 calories.[46]

Estimates of the calorie consumption of Domesday households are presented in Table 5.5. While three assumptions of household size are included, only point estimates of calorie requirements for individual household members have been employed. The main reason, apart from

Table 5.5 The energy requirements of the subsistence household (kilocalories per day)

	(1)	*(2)*	*(3)*
Head	1 900	1 900	1 900
Wife	1 050	1 050	1 050
Children	2 125	2 457	2 783
Older adult	570	750	945
Total	5 645	6 157	6 678
Average	1 254	1 231	1 214

Notes: 1. The following standard has been adopted: adult male 1900; adult female 1500; children 878 (based upon a sibling group including one infant, one young child and one older child); older adult 1500.
2. Household sizes: (1) = 4.5; (2) = 5.0; (3) = 5.5.
3. The ratio between the average consumption of a household and that of adult males is 65%, compared with that of two-thirds in Roman Egypt. See Clark, *The Conditions of Economic Progress*, pp. 660–1, and Goldsmith, 'Roman Empire', *Review of Income and Wealth*, p. 266.

Source: See text.

producing an unnecessarily complicated table, is that an overall sensitivity test is applied when the final national income estimate is calculated. This test suggests that growth rates are highly insensitive to quite large changes in our assumptions.

Estimates of calorie intake can be converted into quantities, and hence values, of food consumed. First, as there are some extant data on grain prices, I have assumed that all the calorie requirements are satisfied by the consumption of grain. This is not a realistic assumption because, owing to the absence of the essential amino acid lysine, deficiencies will quickly emerge in a population if grain is used to supply more than 80 per cent of human food requirements. The assumption made here is only that the cost of providing additional foods that are rich in lysine, such as legumes, is similar to the estimated expenditure on grain. Prior to the thirteenth century the diet of peasants consisted mainly of a pottage containing barley and/or oats (and sometimes peas and beans), bread made from barley, and on much rarer occasions some cheese, milk, ale, and a little salted fish. Food items such as wheat and meat were grown for the consumption of manorial lords and for trade.[47] Secondly, calorie requirements can be converted into grain consumption at the rate of 100 grams of barley per 360 calories.

Thirdly, the estimates of average household consumption of grain have been valued using the available prices for barley and oats.[48] The use of market prices to value subsistence output can be justified on the grounds that 'prices' in the subsistence and demesne sectors shown in Figure 5.4 (on page 203) will be equated by flows of resources between the two sectors. No attempt has been made to include the value of processing grain, because this was a 'household' rather than a 'market' activity. Peasant grain was stored as such and was ground, using household grinding stones (or 'handmills'), only when a meal was being prepared. The roughly ground grain was then included in a pottage, and occasionally used to bake bread for the family. I have treated this activity as meal preparation, which is consistent with the national income estimate of 1688. Only the lord's grain was ground on a large scale for the market, using the relatively recent watermill technology. Fourthly, as shown in Table 5.6 (on page 190), the estimated average household consumption values were applied to the number of unfree peasant and slave households (232 107) recorded in Domesday Book. It is essential to realise that the subsistence sector does not include free peasants because their annual values are recorded in the demesne income taken from Domesday Book.

Figure 5.2 A reconstructed Anglo-Saxon village

Note: An Anglo-Saxon village, West Stow, Suffolk with the Church of St Mary Magdalene, Rothwell, Lincolnshire.
Source: PRO, *Domesday 1086–1986: Exhibition Guide* (London, HMSO, 1986) p. 16.

To cover the cost of clothing, housing, and heating – all of which were provided from materials on the manor (wool and hides; wattle and daub or split timber; and peat or firewood) during downturns in seasonal activity, when the opportunity cost of peasant labour reached relatively low levels – a mark-up of 10 per cent, which is probably generous, has been added. Cipolla provides evidence to suggest that in England, as late as the fifteenth century, expenditure by the 'mass of the population' on clothing, textiles, heating, light and rent was only about 13 per cent of total purchases, and concludes that, 'after having bought their food the mass of the people had little left for their wants, no matter how elementary they were'.[49] In the eleventh century, as illustrated in Figure 5.2, the standard of housing was exceedingly poor.

Archaeological evidence suggests that peasant houses, which were 'crude and flimsy structures' of a temporary nature, normally consisted of one small room (say 10 by 12 feet) that was open to the straw-thatched roof, had small unglazed windows, an earthen floor, and a central hearth without a chimney. These structures, which invite comparisons with the poorer parts of rural Africa today, were often extensions of similar shelter provided for the peasants' animals. It is doubtful that the opportunity cost of building and maintaining such structures amounted to more than 4 per cent of the cost of household food consumption. Peasant houses were only very sparsely and crudely furnished and contained a few wooden utensils, together with wood and clay tableware, which were generally produced by the household itself. It is considered unlikely that the majority of peasant households possessed any cloth, bedlinen or window curtains, and most peasants had no more than a single outfit of roughly made clothes: Cipolla tells us that the 'Peasants were always clothed in rags'.[50] Even so, the rough woollen tunic was the most valuable item that the peasants owned, and may have cost in the vicinity of 0.25 pence per annum, or 1.7 per cent of the value of food consumed by an adult male,[51] and they were eagerly acquired by relatives when their owners died.[52] Warmth and light in winter were derived from the hearth fire that was fuelled by peat and/or twigs dug and collected on the manor, and extra warmth was provided by animal skins. Living conditions for the mass of the population in eleventh-century England were, therefore, of an extremely primitive and precarious nature.

The results of these calculations are given in Table 5.6. In order to test whether my estimate of subsistence income per capita would have been sufficient to maintain the unfree peasant labour force in working condition, I have extracted income data from Domesday Book for a large (40 per cent) and representative sample of 'small' free peasants, who were located virtually entirely in the eastern counties. Small free peasants were defined as those who worked their own land with the labour of their families and/or unfree peasants who were at the lower end of the scale (that is, bordars and cottars but not villeins). If small free peasants, who traditionally have been regarded as being on a par with unfree peasants, received per capita incomes not significantly different from the average for unfree peasants, then at the very least we will know that my estimate of subsistence income was indeed sufficient to maintain a working population. Also, as the Domesday Survey was taken during a period of wholesale property confiscation following invasion – including the forced reduction of the free peasantry to a

Table 5.6 Household consumption in the subsistence (or unfree peasant) sector

H/hold size	*H/hold energy requirements (K/cals)*	*Grain consumption per h/hold (qtrs/annum)*	*Value of grain consumption/ h/hold p.a. (shillings)*	*Total value of subsistence food consumption (£)*	*Value of total subsistence income (£)*	*Subsistence income per capita (pence)*
4.5	5645	3.16	3.692	42847	47132	10.8
5.0	6157	3.44	4.019	46642	51306	10.6
5.5	6678	3.74	4.381	50846	55931	10.5

Notes: 1. 100 g of barley provides 360 K/cals (Paul and Southgate, *McCance and Widdowson's The Composition of Foods*).
2. One bushel of barley is equivalent to 22.68 kg.
3. In 1086 there were 232107 households in the subsistence sector.
4. It is assumed (unrealistically) that grains provided 100 per cent of energy requirements (see text).

Source: See text.

bonded status – we can expect that the average income of the peasants was less after confiscation than before.[53] The explanation is that the invaders would only expropriate free peasant land if, after making sufficient land available to the former owner for subsistence (in return for working the balance of the land for its new owner), there was a surplus that could be extracted. Evidence on this is abundant in both Domesday Book and the Anglo-Saxon Chronicle.[54] It should also be borne in mind that 1086 was not a period of equilibrium as the confiscation process had not been completed. In the light of this reasoning we can expect the per capita income of unfree peasants to be even less than that of free peasants.

The data to test these hypotheses are presented in Table 5.7. The average income of small free peasants in my sample of six counties is only 16 per cent higher than that of unfree peasants, and that for Norfolk, which contained the majority (87 per cent) of the free peasants in the sample, only 5 per cent higher than that of unfree peasants. The difference in average income for small free peasants between Norfolk and the entire sample has been generated by a higher average for Essex. It is clear, therefore, that my estimate of subsistence income was definitely sufficient to maintain a working population, and that it appears to approximate the real level of per capita subsistence

Table 5.7 A comparison of average income of small free peasants and unfree peasants, England, 1086

	Number (000s)	*Percentage of total*	*Average income (pence)*
Small free peasants			
Six counties	4.3	39.7	12.1
Norfolk (87% of total)	3.7	28.9	11.1
Unfree peasants (excl. slaves)	197.9	100.0	10.6

Sources: Free peasants – Domesday Book entries (both lay and ecclesiastical) for Norfolk, Essex, Cambridgeshire, Nottinghamshire, Northamptonshire and Derby: Snooks and McDonald Domesday database. Unfree peasants – Table 5.6.

income. A further important test of the reality of my estimate, as will be discussed below, is that GDP per capita of England for 1086 is about the same as that for India in the mid-nineteenth century. This is a sensible and persuasive comparison.

Finally, it is important to compare my 1086 estimate of consumption standards with other estimates for subsequent centuries in England. Bruce Campbell, after an examination of all available evidence, has concluded that the average energy intake for the entire population of London (and England) in 1300 was in the range 2000 to 2500 K/cals per day.[55] In order to make a valid comparison with this 1300 estimate it is necessary to extend my estimate of consumption levels of unfree peasants to include the entire population by adding free peasants and manorial lords; and to take into account any change in average consumption levels between 1086 and 1300. First, by taking the consumption levels of free peasants and manorial lords into account (Table 5.9 on page 200), the average level of energy consumption increases from 1230 to 1500 K/cals per day. Second, the growth rates presented in Chapter 7 (see Tables 7.5 and 7.6 on pages 248 and 250) suggest that between 1086 and 1300, per capita income increased by approximately 92 per cent and average consumption by about 48 per cent, which increases the energy consumption estimate from 1500 to 2200 K/cals per day. This lies midway in the Campbell range. Hence my estimate of calorie consumption in 1086 is equivalent to that of other scholars for 1300.

Manorial income – omitted counties

Domesday Book excludes four counties to the north of Yorkshire – Cumberland, Westmorland, Northumberland, and Durham – and the entries for Lancashire are incomplete. In order to obtain an estimate of national income for the whole of England, for comparison with later estimates for the entire country, it is necessary to calculate manorial income for the four northern counties and for Lancashire. A reworking of the estimates by Darby suggests that the population of Lancashire ranged from 8100 to 9900 and that of the four northern counties from 22 500 to 27 500 (see Table 5.2 on page 178), depending upon the assumed household size.[56] Income estimates were obtained by applying the recorded manorial income per capita to these population estimates. Some might wish to argue that the average per capita income for those counties south of the River Tees is too high for the four northern counties. The extent to which this is so will determine the degree to which my growth rates have been underestimated. The summary results are given in Table 5.8 (on page 194) and the detailed results in the Appendix (Table A.1 on page 278).

Burghal income

Domesday Book does not *systematically* record the population or economic activities of boroughs or towns, except where these are concerned with manorial issues. As I have speculated elsewhere, this could be due not only to the feudal purpose of the survey[57] – as M. M. Postan noted, medieval towns 'were non-feudal islands in the feudal seas'[58] – but also to a physiocratic-type view of the production of wealth. For one or both of these reasons towns, which held no more than 8 per cent of the population of England, seemed to be of little interest to William and his Exchequer when compiling Domesday Book. Despite this partial omission in Domesday Book it is possible to reconstruct the urban population of England. The Conqueror's survey makes reference to 112 towns scattered throughout the country, and to a variety of resources they contained, such as burgesses, houses, churches, mints, land, ploughteams and so on. From this incomplete picture, together with other sources of information (for example, law codes, guild regulations, and coins), it is possible to estimate the size of individual towns with a reasonable degree of accuracy. The most careful estimate suggests that the urban population of England in 1086 was about 120 000 people, or 8 per cent of the total population.[59]

The value of net production in Domesday towns can be estimated by applying manorial income per capita to the urban population estimates. This method is based upon the quite reasonable assumption that the marginal *and* average returns of manor and borough will tend to equality, at least within a given locality. This assumption is reasonable because individual wealth was not highly concentrated in terms of either scale or location. Typically, a tenant-in-chief had a central manor around which some of his properties were clustered, with the rest of his relatively small-scale holdings scattered widely throughout England.[60] In addition, tenants-in-chief were accustomed to travelling regularly between their holdings in town and country (as I have argued elsewhere),[61] mainly because it was more efficient for a household to move around England than to transport bulky commodities with a low value per unit of volume. It was also an effective way of checking on the activities of their reeves. In these circumstances, manorial lords would have been keenly aware of the returns from their various small-scale holdings. In all probability, if the returns on individual rural and urban holdings were seriously jolted out of alignment, there would have been a resulting shift of resources. This process would have produced a tendency to equality of returns not only between small individual holdings (which can be regarded as an approximate way of capturing marginal returns), but also between aggregated (*average*) rural and urban holdings *in a given locality* (where other things were constant). In effect, Norman tenants-in-chief held a 'portfolio' of small and scattered urban and rural assets, which would have been an effective hedge against uncertainty in a world where invasion, disease, and natural disaster were ever present. Indeed, they were accustomed to shifting assets around the country when disaster struck, or was expected to strike in the case of anticipated invasion, as happened quite regularly. Domesday Book records, for example, many cases of manors that were 'waste' as a result of natural and man-made catastrophes. Of course, in a dynamic economy, complete equality – or equilibrium – would never be achieved. But if there were a significant and persistent difference between rural and urban returns, then we could anticipate that per capita income would be higher in the country because that is not only where England's wealth was generated and recorded, but also where the rich landowners lived and held their assets. Surely it is for this reason that the urban population was so small, and that the rate of urbanisation during the following two centuries was relatively slow. It took 600 years to double the relative size of the urban sector. In any case, urban income was only a very small proportion of total income.

The estimated total for burghal income is shown in Table 5.8. This has been calculated by employing the per capita manorial income figure (1.79 shillings) associated with an average household size of 5.0 people which, as argued above, is the most plausible assumption underlying the estimate of population. The Appendix shows how per capita income varies with different household size assumptions, and provides a discussion of the implications of the variations.

Table 5.8 The national income of feudal England, 1086

	£	s.
1. National income		
1.1 Manorial income (£)		
1.1.1 Recorded counties		
demesne economy	71 573	
subsistence economy	51 306	
1.1.2 Omitted counties	3 034	
1.2 Burghal income	10 708	
1.3 Total	136 621	
2. Per capita income (shillings)		1.785

Notes: Non-recorded income estimated on the assumption that the average household size was 5.0 people, and that slaves were recorded in the same way as the rest of the population (that is, as household heads).
Sources and methods: See text.

The national income of feudal England, and a comparison over two millennia

The resulting estimate of England's national income for 1086, shown in Table 5.8, is £136 621. As the population in the survey year was, on the most plausible set of assumptions,[62] 1 531 000, the per capita income of Domesday England was 1.79 shillings and income per household was 8.58 shillings. Alternative estimates of national income, ranging from £133 133 to £139 507 (a variation of only 4.9 per cent) using different population totals are given in Table A.1 in the Appendix. Not surprisingly, the distribution of income per household (that is, recorded person) varied between counties. Figure 5.3, which illustrates the geographical distribution of demesne income per rural household, shows that the highest returns in 1086 (just as in 1991) were achieved in the south-eastern part of England, particularly in the old Anglo-Saxon kingdom of Wessex.

A question of major interest is how the performance of feudal England compares with that of other economic systems through time. Differential levels of average income can be used as a rough

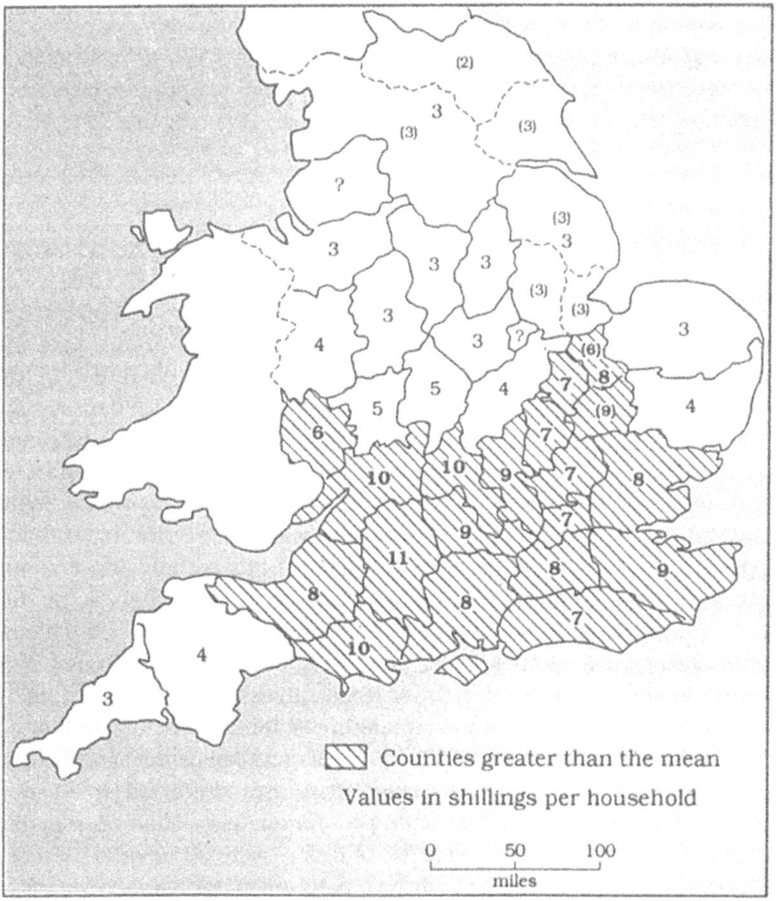

Figure 5.3 Geographical distribution of income per household, England, 1086

Source: Darby, *Domesday England*, p. 229.

index. The available evidence suggests that GDP per capita in the English feudal system (1086) was about half that in the early Roman Empire (AD 20); was about 18 per cent of that of the English mercantile capitalist system (1688); and was on a par with that of India in the mid-nineteenth century.[63] In other words, the per capita income of the early Roman Empire was less than half that of England in 1688, but more than double that of England in 1086 and India in the 1850s. Hence the European feudal economic system provided an average income substantially lower than that of the ancient world at the beginning of the Christian era, but one that was on a par with the average income provided by post-industrial Third-World economies in southern Asia some 800 years later. This rough calibration provides a guide to the ups and downs of human society over the last 2000 years.

Sensitivity analysis

An attempt has been made in this chapter to provide a realistic estimate of national income of feudal England. The fact that GDP per capita in Domesday England was similar to that in India in the mid-nineteenth century seems sensible, but how robust are my estimates and how do they stand up to sensitivity analysis? In the first place, my income estimates do appear robust. Whenever the choice of assumptions involved in the estimation of subsistence income was not clear I adopted the one implying the *lowest* rate of growth of per capita income because it was clear that this would be the contentious issue. This is also true of my selection of price indexes and population estimates. A good illustration of this method concerns my estimate of population for 1086, which at 1.53 million is on the conservative side. Some authors have argued for a larger estimate of population, ranging from 1.75 to 2.25 million;[64] an argument that possibly gains some support from recent upward revisions for 1300.[65] Even a modest increase in the population estimate from 1.53 million to 1.67 million would reduce GDP per capita and hence increase the rate of growth of per capita income from the preferred 0.29 to a significantly higher 0.32 per cent per annum (see Table A.1 in the Appendix). The growth rate rises directly with population because demesne income is a large constant component of national income, which reduces GDP per capita within the population range suggested above. The population effect, therefore, is more important than any marginal revision in my 1086 national income estimate (particularly of the peasant economy) that readers may be tempted to suggest.

In the second place, it can be shown that my growth-rate estimates are insensitive to changes in the estimating assumptions. If, for example, the size of the unfree peasant economy was increased by 50 per cent – placing the average income of unfree peasants substantially (35 per cent) above that of small free peasants – the rate of economic growth would decline only from 0.29 to 0.26 per cent per annum, which exactly balances the population effect. A rate of growth of 0.29 per cent per annum would appear, therefore, to be a realistic and robust estimate. Even if it were decided to double the size of the subsistence sector, the growth rate would only decline from 0.29 to 0.23 per cent per annum, and then only in the absence of the population effect. And this would be equivalent to increasing the calorie intake of the average family for *every* day of *every* year from 1231 to 2462, or of an adult male from 1900 to 3800, and to raising the average income of unfree peasants some 80 per cent above that of small free peasants – an absolutely fantastic suggestion. Fine-tuning the national income estimates, therefore, will have only a marginal influence upon the growth rates presented in Chapter 7, and no influence upon the overall conclusions of this study.

RECONSTRUCTING ENGLISH FEUDALISM

Using the aggregated data that have been reconstructed here from Domesday Book, it is possible to demonstrate how England's feudal economic system functioned. We can now outline the structure of the Domesday economy, and show how it operated at the macroeconomic level. (It will be shown in Chapter 6 how the feudal system operated at the microeconomic level.)[66]

The structure of the Domesday economy

Just over half (53.7 per cent) of GDP in the Domesday economy was generated by the demesne sector, which produced perishable commodities for the consumption of feudal tenants-in-chief (including free peasants), and for sale through markets in vills, boroughs, city-ports, and in Europe. The dependent sector, in which the subsistence requirements of the bonded peasants were barely met, generated a further 38.5 per cent of GDP, and the small urban sector – which provided manufactured products, a range of public and private services, and facilitated the export and import of a range of commodities – produced a further 7.8 per cent of GDP. These are the bare facts of the feudal economy, which were revealed in Table 5.8 for the first time.

A central issue concerning the English feudal economy is the extent of the market both for commodities and factors of production. How far had commercialisation proceeded by 1086? We know from other sources that there was extensive internal and international trade in this period. There is, for example, evidence of an active trade in textiles, metals (iron and lead), salt, horses, and agricultural goods such as dairy products and grain. Domesday Book tells us of salt-making, a commodity vital for communities that needed to preserve fish and meat for the winter, which was undertaken at a number of inland centres (including Worcestershire and Cheshire) and at many places along the east and south coasts. A glimpse of what must have been a nationwide system of trade is provided by Domesday entries recording manorial involvement in far distant salt-making centres such as Droitwich, together with tolls scaled according to distance that were imposed upon purchasers.[67] Another essential industry that provided the basis for a thriving trade was iron-making. The products of this industry, which included ploughshares, farm implements, horseshoes, and weaponry, were required on every manor. Centres of ironworking are mentioned in Domesday Book for Cheshire, Devonshire, Hampshire, Herefordshire, Lincolnshire, Northamptonshire, Surrey, Sussex, Warwickshire, Wiltshire, and Yorkshire. These numerous centres would have purchased from other regions those raw materials not found locally, and would have sold raw iron and iron products to manors throughout England and, as the reputation of English ironworkers increased, to Europe as well. Other metal crafts included leadworking in Derbyshire, the output of which was used in the building industry (particularly church and manor house roofing), and for making salt vats and other similar containers. These commodities would also have been an important item of internal trade, as may have been tin and tin products from Cornwall despite their omission from Domesday Book.[68]

It would appear that the less sophisticated products were produced on the manor (with the exception of glass, which was largely imported into Anglo-Norman England), while finer items were purchased from specialised craftsmen in the major boroughs of England and Europe. Sufficient evidence is available, however, to demonstrate that extensive trade existed in industrial and agricultural commodities between the manors of England, and that this trade was facilitated by a network of boroughs throughout the country. The widespread establishment of mints and the resulting supply of adequate and acceptable coin provided the necessary medium for the development of an extensive system of internal trade.

By the time of the Conquest, European trade, undertaken at first by Frisian and later by Scandinavian seamen and merchants, linked England with all parts of the Continent. The late Anglo-Saxon economy had trading connections with Scandinavia, Flanders (and from there to Cologne and Bruges), southern Europe and the Mediterranean.[69] In response to this commerce with Europe, a number of English coastal towns, such as London, Southampton, Dover, Hull, and Ipswich, developed into thriving ports, occupied by a rapidly emerging merchant class in the tenth and eleventh centuries. As the king imposed tolls upon ships entering these ports, he was keen to encourage this development. The main exports from England in the mid-eleventh century included both craft and agricultural products. Indeed, fine English craftwork, in the form of metalwork (particularly silver), embroidery, fine textiles, and manuscripts, appears to have been in considerable demand throughout Europe. In addition, there were exports of agricultural products, including grain, cheese, butter, oil, honey, and salt. Imports, on the other hand, appear to have been mainly luxury goods that could not be produced on the manor or in the borough. These imports included silks, finely dyed textiles and garments, gold and precious stones, glass, wine, oil, ivory, tin, bronze, copper, and sulphur. Other less exotic imports included military equipment, timber, furs, hides, and fish.[70]

We also know that the manorial lords needed to market their surplus in order to provide cash to meet the taxes, both feudal and national (amounting to about one-third of the annual values),[71] to purchase military equipment and warhorses, and to satisfy their desire for elegant building materials (glass) and luxuries, fittings, clothing, and investment in productive rural and urban assets. Yet to gauge how extensive the market really was in 1086, we must be willing to make a number of plausible, but largely unverifiable, assumptions in order to adjust the national income estimates given in Table 5.8.

An estimate of the value of net demesne output (that is, after the deduction of seed, fodder and so on) that entered the market is presented in Table 5.9. While the assumptions are plausible and lie within empirically-determined boundaries, they are not precise. Accordingly, the estimate of the degree of commercialisation is fairly rough. Nevertheless it may not be very far from the mark to say that

Table 5.9 Estimate of demesne surplus entering the market sector

	£
Demesne income	73 340
minus consumption of perishables by free peasants	−16 328
minus consumption of perishables by tenants-in-chief	−12 878
Traded surplus	44 134

Notes: 1. The per capita consumption of perishables by all (not small) free peasants is assumed to be twice that of unfree peasants. Hence out of the average free peasant's income of 56.4 pence (Norfolk), 21.24 pence were consumed, about 18 pence paid in taxes, leaving 17.2 pence for other expenditure including land improvement.
2. The per capita consumption (perishables only) of tenants-in-chief and sub-tenants is assumed to be greater than that of free peasants by a factor of four (and unfree peasants by a factor of eight).
3. Assumed that all perishable consumption met from own holdings.
4. Demesne income includes omitted counties.
5. Taxes had to be paid in coin, not kind, and therefore output had to be sold.

Source: Table 5.8.

about 60 per cent of net demesne output in 1086 was marketed. Also, as primary produce to the value of £44 134 passed through an urban sector which contributed only £10 708 to GDP, the bulk (possibly as high as three-quarters, even if we assume that *all* urban income was spent on domestic rural goods) of this marketed surplus (or 24 per cent of GDP) must have been exported.[72] Certainly, these conclusions are consistent with the indirect evidence of market activity discussed above. All this evidence points to the conclusion that the feudal economy of England was substantially open to market influences and must have been capable of responding to them.

It is now possible to reorganise the national accounts to determine the relative importance of the market and subsistence/sustenance sectors. Table 5.10 suggests that 40 per cent of the economy was involved in market activities and 60 per cent in subsistence. In turn, the market sector was dominated by manorial agriculture, amounting to one-third of GDP, and the subsistence sector was dominated by the dependent peasant sector, amounting to almost 40 per cent of GDP. This challenges the conventional wisdom which insists upon only a very limited role for market forces at this time. Wilhelm Abel, for example, claims:

Table 5.10 Market and subsistence sectors in the feudal economy of England, 1086

	£	%	£	%
Market				
rural	44 134	32.3		
urban	10 708	7.8		
Total			54 842	40.1
Subsistence/sustenance				
tenants-in-chief	12 878	9.4		
free peasants	16 328	12.0		
unfree peasants and slaves	52 573	38.5		
Total			81 779	59.9
Totals			136 621	100.0

Notes: Rural income, both market and subsistence, includes omitted counties on a *pro rata* basis; all urban income has been treated as market income.
Sources: Calculated from Tables 5.8 and 5.9.

> Around the first millennium A.D. trade in central and western Europe still consisted almost entirely in the exchange of a few, usually inessential commodities within a very limited area . . . In inland districts . . . the economy of the manors and farms was almost exclusively of the subsistence type . . . As yet prices and wages had little importance. Such prices and wages as have come down to us often leave us uncertain as to whether they were still the symbols of a fixed system of values, rooted in law and tradition, or whether they had already become values of exchange, determined by the scarcity of goods.[73]

There can be little doubt from my estimates that prices, at least, were indeed the result of market transactions. My estimate of the size of the market sector in the feudal economy of England can be placed in a wider historical context by comparison with modern Third-World economies and with the ancient world. In the mid-twentieth century, the market economy accounted for about 50 per cent of GDP in Africa and 60 per cent of household consumption in India.[74] While there is no direct evidence for the Roman Empire at the beginning of the Christian era, R. W. Goldsmith is prepared to hazard an informed guess, that 'the monetisation ratio of national product is unlikely to have been as high as one-half'.[75] If this is so, it would appear that the relative size of the market sector in the English feudal economy was only marginally less than that in both the early Roman Empire and Third-World economies in the mid-twentieth century. This has important implications that will be explored in the following chapters.

A macroeconomic model of the feudal system

By employing the above national accounting data it is possible to construct a simple macroeconomic model, in diagrammatic terms, of the feudal system in England in the late eleventh century. Figure 5.4 illustrates the three main elements in the feudal economic system: the feudal state; the manorial economy; and the market economy.[76] The manorial economy is based upon an unusual interaction (in the light of modern experience) between the demesne sector and the dependent peasant sector, involving an exchange of land and capital for labour, and of military service for justice and protection. Within both these sectors there is a further interchange between the household and the farm, consisting of an exchange of labour for sustenance. The market economy is based upon a relationship between the demesne sector and the urban sector (with further interactions with the 'international' economy), with an exchange of primary products for manufactured production, military equipment, horses, luxury goods and so on, and, to a lesser degree, an exchange of capital and labour for interest and rent (labour is a scarce resource). Within the urban sector there is an exchange of labour and wages between the household and workshop/commercial/service sectors. Finally there is an exchange between the feudal state and both the manorial and market economies, involving property rights, jurisdiction, security and so on, on the one hand, and taxes, dues, and military service on the other.

An interesting part of this economic system is the relationship between the demesne sector and the dependent peasant sector. While both sectors were based upon the production of agricultural products, they were, in fact, segmented markets. There was a flow of resources – land in one direction, labour in the other – but not of commodities or cash. Nevertheless, commodity 'prices' between the two sectors would have been equilibrated by the exchange of factors of production. Also, these two sectors tended to specialise: the demesne in commercial

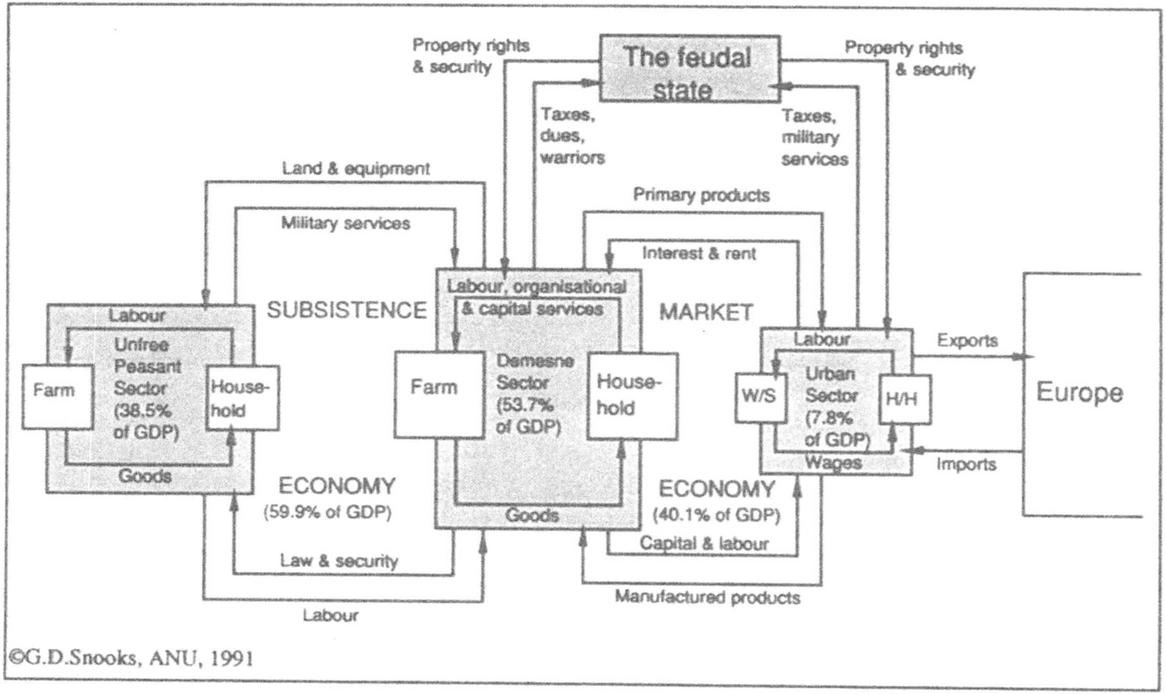

Figure 5.4 **A diagrammatic model of the English feudal economy, 1086**

Source: Constructed from data in Tables 5.8 and 5.10.

products (wheat, wine, and wool), although it also produced subsistence crops; and the dependent sector in subsistence crops (barley, oats, and other non-traded food).

But how did this system actually work? What were the incentives, punishments, and institutions that enabled the ruling elite, which constituted only about 0.5 per cent of the entire population, to so dominate this society that they could keep the vast majority at subsistence level and expropriate the entire surplus for themselves? Feudal systems have flourished during periods of insecurity: as H. R. Loyn has written, 'In an age of peril it was indeed natural for men to seek lords, and lords to seek men'.[77] While at first freedom may be traded to gain security, once the political and economic system of feudalism finds its fullest expression, as in England after the Conquest, there is little room for choice. Anglo-Norman society was a highly centralised form of feudalism, in which tenants-in-chief held their land directly from the king in return for military (or 'knight') service, taxes, and feudal dues. Through a process of subinfeudation, this land filtered down the feudal hierarchy, and at each level there was an exchange of property for promises (the swearing of oaths) of loyalty and military and other services. The bonding agent – in all senses of the word – in this hierarchical system was the military strength of a highly successful warlord in a period of wars and threats of wars. Royal control was maintained by dispersing the territorial power of the barons through the widespread scattering of estates, and by close surveillance through the requirement of regular attendance at court, through an efficient system of local, or shire government developed by the Anglo-Saxons, through a network of spies, and through severe punishments involving confiscation of land, mutilation, imprisonment, death, and excommunication.

Control of the bulk of the population at the bottom of the feudal hierarchy – the dependent peasants – was achieved through monopsony agreement between barons, manorial lords, and clergy. Labour was the scarce factor in a resource-rich country such as England in 1086, and the best way for the ruling elite to maximise their surpluses was to agree not to compete for labour but rather to extract rent from them by terror in this life and the next – the dual threat of the castle and the cathedral.[78] As there was nowhere to go – bonded peasants would either be returned if they attempted to flee their manorial lord or forced to live a precariously short existence in the swamps or forests as outlaws – those at the bottom of the feudal hierarchy just had to adjust to their lot. And to see that they did, they were closely supervised at their work by reeves and the lower clergy, and were abandoned by the king to the jurisdiction of manorial lords who were also warriors. Positive incentives took the form of undertaking their mandatory work on the demesne as quickly and as effectively as possible so as to be allowed to work on their own plots of land to ensure survival. The narrow margin between life and death was incentive enough. It was not until the decline of the feudal system from the thirteenth century onwards, owing to a decline in the predatory and aggressive competition between warlords that had enabled this system to flourish, that the peasantry began to acquire sufficient bargaining power to capture a greater proportion of their marginal product.

Unlike the dependent peasants, the manorial lords were well situated to respond to market incentives. Manorial surpluses sold through the market could be employed not only to meet the heavy obligations to the king and other feudal lords, but also to acquire wealth that could be translated into political power. The essence of feudalism is the competition between warlords. As William forbade warring between barons in England, but not in Normandy, the only form of competition possible was economic and political. Wealth and political influence in England could be enhanced by effective and innovative management of estates, because growth is the most effective way to maximise income and wealth over one's lifetime.

Both the incentives and the opportunities in terms of surplus income existed for manorial lords to act entrepreneurially. The question is, did they? The usual assertion is that the ruling elite wasted these opportunities through riotous living. We need to test this assertion. Accordingly, an attempt will be made in Chapter 6 to examine the issue of economic rationality, by asking the central question: was medieval man economic man? Further, while England's feudal economy was a brutally inequitable system, it was capable of expropriating and mobilising a relatively large surplus that could be used for productive investments necessary for the generation of growth. Had the system been more equitable it is doubtful that any measure of growth would have been possible. In Chapter 7 an attempt will be made to see if this potentiality was, in fact, translated into economic growth.

6 Human Motivation Throughout Time

Consent or custom

(John Milton, 1667)

ECONOMIC MAN OR MORAL MAN?

The mere existence of market forces in an economic system may not be regarded as sufficient evidence that economic agents respond effectively to the material incentives these markets provide. We have been told repeatedly by historians and by economists in the classical tradition that the ruling elite in the Middle Ages preferred to ignore market signals by consuming rather than investing the surpluses they extracted from the peasantry. Indeed, we are expected to believe that medieval man, unlike his modern counterpart, was moral man rather than economic man.[1] Some have challenged this conventional wisdom – such as Marshall in his debate with Cunningham (see Chapter 2), McCloskey in his discussion of medieval agriculture, and North and Thomas in their speculations about longrun institutional evolution[2] – but, owing to a lack of hard evidence, their ideas have made little headway. Yet the evidence is available, in Domesday Book, to test this issue directly. It enables us to answer the fundamental question: were the ruling elite of Domesday England prisoners of custom, or were they primarily motivated by individual material self-interest? Were, in other words, feudal barons economically rational?

No other question can raise as much passion and misunderstanding amongst scholars, particularly historians, than the question of economic rationality. To claim that human societies and even individuals – the second follows from the first – are driven by an overwhelming desire to maximise their material self-interests, is taken by many to be a denial of higher human aspirations or a denial of the idea that ideas *per se* determine human actions.[3] The concept of economic rationality does indeed imply that human society is primarily driven by individuals and groups struggling to maximise their own material returns, either as an end in itself or as the best way to achieve other closely related objectives such as power. This does not deny, however, that some individuals, or social groups, are capable of pursuing 'higher' values to the detriment of their material standing, just that it involves only a small minority of the total population. And it certainly does not deny that wider cultural values are adopted when they are not in serious conflict with material standards of living. But it does imply, when there is a serious conflict between material self interest and cultural values, that this conflict will generally be resolved in the interests of economic man.

Economic rationality can be viewed from a number of angles. The classical economists saw it in much the same terms as described above,[4] whereas the neoclassical economists have taken a narrower and more instrumentalist point of view. In the neoclassical tradition, economic agents have well-established preferences, access to complete information, and amazing computing abilities.[5] Faced with the need to make a decision, economic man gathers all the necessary information and effortlessly computes the outcome that will provide him with the greatest satisfaction in a benefit-cost sense. Neoclassical economic man, therefore, attempts to maximise his objective function subject to constraints. This simple view of human decision-making can be dealt with in the neoclassical framework in a more complex and sophisticated (although not necessarily realistic) way by introducing ideas about risk, uncertainty, information costs, and interactions with other agents (game theory), but the essentials of *homo economicus* remain basically unchanged. Clearly, the assumptions underlying this neoclassical version of economic man have been taken to an unrealistic extreme, and as a reaction alternative models such as 'satisficing' – opting for a 'good' outcome which is not necessarily the 'best'[6] – and bounded rationality – which takes into account the cognitive limits of the decision-maker – have been developed.[7] But as these alternative models do not lend themselves to elegant mathematical solutions – being forced instead to employ less satisfying simulation techniques – they have not made much progress.

No one would argue that the neoclassical ideal type of decision-maker is completely realistic. But then this is not an essential requirement for historical work, provided economic theory is regarded as a box of tools for the exploration of reality rather than a description of that reality. It all depends just how seriously economic theory is taken for its own sake. The extent to which assumptions about complete information and unlimited computing abilities are not true is the extent to which the precision of the predicted outcomes of neoclassical theory will be affected. This need not be a major problem if we take the essentially pragmatic view that economic tools with less than razor sharpness are still useful in understanding reality. This may worry the gameplayer who lives in a very precise world, but not the realist, who is accustomed to viewing the world 'through a glass darkly'. In any case, the otherworldliness of neoclassical economic man and the imprecision of deductive theory cannot be taken to deny that in reality economic agents may attempt to maximise their material interests in an imperfect world, or that mistakes are made and outcomes are often different from expectations. It would be dangerous, however, to mistake errors of fact and judgement for irrationality. But it would be equally foolish to deny that irrationality exists and that governments and even economic systems fail as a result – the Thatcher government is an example of the former and Eastern Europe is an example of the latter. What we need to do in this study is to employ the evidence of medieval England to resolve the debate about competing behavioural models in history. This is at least in the spirit, if not the letter, of Herbert Simon's claim that 'the model of human rationality must be derived from detailed and systematic empirical study of human decision-making behaviour in laboratory and real-world situations'.[8]

The real issue, it will be argued, is not about economic man versus moral man, but about how best to achieve the maximisation of material self interest and about who is doing the maximising. Debate on this issue often confuses means with ends. The ends are the maximisation of material self interest, and the means are the creation of competitive or protective environments or, more dramatically, of chaos or order. There appears to be a tendency in arguments on this issue to identify social order or protective environments – which has often been identified with English feudal society – with the rule of custom. In reality, different individuals, groups, or societies will, at a point in time, choose either means to achieve their ends. Much more will depend upon the competitive environment in which individual societies find themselves. If societies are geographically isolated, as were Aboriginal Australia and China before the Industrial Revolution, it may be possible to maximise immediate material returns by establishing internal order and consensus (although the longrun cost of isolation is the inevitability of economic take-over). But if societies are subject to constant external pressure, as were those in Europe over long periods of time, then their economic objectives can only be achieved through more aggressive internal (and external) relationships.

In the more open European situation, established rent seekers, for example, may find it expedient to restrict competition between themselves in certain spheres (such as the use of peasant labour by feudal tenants-in-chief) but not others (such as in the commodity market or on the battlefield or at the king's court). Those outside the system will, however, favour aggressive competition so as to break into the closed inner circle, and will be prepared to disrupt the existing order through the creation of chaos. But as soon as the outsiders have succeeded in riding the whirlwind they will want a return to order and the protection from outsiders of what they have recently gained. This oscillation is not between moral and economic man but between the alternative strategies of *homo economicus*, in open and closed societies. Such strategies are not restricted to feudal society. In modern society, vested interests, particularly those that are not confident about their competitive abilities, favour protection of the existing order so as to maximise their material returns through a combination of productive pursuits and rent seeking, at the expense of the consumer who is supporting the tariffs and subsidies. Outsiders, on the other hand, will attempt to disrupt the existing protective order by persuading governments to remove protective devices and barriers to entry, so as to promote their own material interests. Once again the struggle is not between economic man and moral man, but between different groups of economic men employing different maximising strategies. In the final analysis, a healthy economy will require a balance between the forces of competition/chaos and protection/order: an excessively aggressive society will eventually tear itself to pieces, even though in the shortrun it may generate impressive growth; and, conversely, an excessively protected society will eventually become moribund, even though at the beginning it may have generated acceptable economic progress. It is a matter of order or chaos, rather than consent or custom.

ECONOMIC MAN IN TIME: FEUDAL ENGLAND

It is significant that much of the surviving documentation from medieval and ancient societies, dating as far back as the Babylonian Empire of some 4000 years ago, is in the form of financial records. From the detailed and precise way in which financial data were recorded in distant civilisations,[9] there can be little doubt that material considerations have always been of paramount interest. Certainly there is no more significant document in medieval Europe than Domesday Book, the compilation of which involved, in addition to all landholders in Anglo-Norman England, the active participation of at least 10 000 temporary public servants (or 0.7 per cent of the population) throughout much of the year 1087. This large and costly undertaking was not the product of idle curiosity, or of a society that took economic matters lightly. But does this mean that feudal society was dominated by material self interest? Only an analysis of the data in Domesday Book will provide the answer.

In an attempt to distinguish between the competing behavioural assumptions of economic man and moral man I have focused upon three key issues that have been used by historians to support the hypothesis that medieval man was moral man. These issues include the system of taxation; the economic role of manorial lords; and the process of manorial production. Of these, the first is an important macroeconomic issue which affects the way we view the role of the state in a feudal system, while the other two are microeconomic issues concerned with the entrepreneurial and organisational abilities of medieval man. The conventional wisdom views the state as an economically irrational institution imposing its power in an arbitrary and wilful way, and views manorial lords as non-economic men who had their estates managed according to the time-honoured communal practices of 'custom', and who frivolously consumed their surpluses. It is gratifying that, since the publication of *Domesday Economy*, at least one economic historian of the medieval period – indeed a reviewer of the book – has kindly taken it upon himself to promote my argument about the economic rationality of the medieval English economy.[10]

This interpretation by medievalists and classical economists has cast a long shadow over historical scholarship. Even historians of the modern world have built it into their explanations of the emergence of capitalism. One influential thesis is that 'Economic development in its European form required above all freedom from *arbitrary* political acts concerning private property'.[11] This interpretation is based upon the notion that feudal society and economy were dominated by custom and the arbitrary intervention of feudal lords. It is claimed that only once the 'old arbitrarinesses' were broken down and once 'risk was cut and custom eroded', could a market economy, and hence growth of per capita income, emerge. The fundamental issue, therefore, concerning not only the medieval economic system but also the explanation of the emergence of capitalism, is whether Domesday England was ruled by custom and arbitrary decree or by individual initiative and rational calculation.

The system of taxation

It is generally thought that taxes on wealth have a very long history, but that taxes on income are relatively recent. As one commentator has said: 'In the historical evolution of government finance, the income tax is relatively novel. It is difficult to find any evidence of a serious national income tax being used until the end of the eighteenth century.'[12] This eighteenth-century tax, imposed by William Pitt in 1799, was used to finance the war with Napoleon and was removed as soon as the war had been won. While there is no doubt concerning the longevity of taxes on wealth, which were commonly used in the ancient world of 4000 years ago,[13] there is good evidence to show that a national income tax was imposed in England as early as 1086 and probably even in 991 when danegeld was first introduced by Ethelred the Unready.

The geld

At the time of the Domesday survey, public revenue was raised from a variety of feudal and non-feudal sources. This was required by the king in addition to feudal services such as knight service and attendance at Court. Feudal revenue took the form of aids (arising from marriages, knighthoods, and wardships) and reliefs (paid on the inheritance of a fief), while non-feudal funds came from the royal demesne, various local and national courts, rights over boroughs and minting, and finally the geld.

In the provision of total public revenue the geld played an important part. After an examination of the rights and obligations of Norman barons, S. Painter concluded: 'The most important public burden of the Norman period was the tax known as danegeld.'[14] He showed that in 1162, the last year in which the geld was levied, an amount of £3132 was collected, which compared more than favourably with the £2408 raised from 'the most profitable feudal aid of Henry's reign'. Postan also claimed that at this time the geld contributed between a quarter and a third of the king's annual revenue.[15] The relative size of danegeld can be reconstructed from the entries of Essex: the rate struck in 1083–84, of six shillings to the hide, implies that the average manor was required to part with approximately 16 per cent of its annual revenue to meet this impost. The geld, therefore, was both an important source of royal revenue and a considerable burden on landholders.

At the beginning, geld was levied on a national basis, as an extraordinary tax to bribe raiding parties of Danes to direct their unwanted attentions elsewhere. This impost, initially known as danegeld, was first levied in 991 by King Ethelred and, until 1012, was collected irregularly to meet the Danish threat. From 1012 it became an annual tax, and was employed to finance both a standing army (known as royal housecarles) and a small fleet of warships, until it was discontinued by Edward the Confessor in 1051. At this time the tax was known as heregeld (or army geld). The Conqueror may have revived the geld as early as 1066–67, but little is known about its incidence until the end of his reign, by which time it appears to have become an annual tax once more.[16] From this time, more information is available. In 1083–84, for example, the geld was levied at the rate of six shillings per hide (the unit of assessment); in 1096 at four shillings per hide; during Henry I's reign at two shillings per hide (levied annually); and during Henry II's reign at two shillings per hide for 1156 and 1162. Although the geld, as such, was not levied after 1162, owing to the opposition it finally aroused, both Richard and John attempted to raise similar taxes on five occasions (1194, 1198, 1200, 1217, and 1220) based upon hides and even actual ploughlands. But renewed opposition from the king's barons meant that even these irregular levies finally ceased in 1220.[17]

The traditional interpretation

For the past hundred years the geld has been interpreted as an 'artificial' taxation system. Traditionally scholars have argued that it resulted from the exercise of arbitrary power by the monarchy, which refused to take into account the capacity of landholders to pay this impost. This view was first put forward by J.H. Round almost a century ago, when he referred to the record of tax assessments in Domesday Book as 'a vast system of artificial hidation' – a system which he named the 'five-hide unit'. His conclusion was based upon an examination of no more than a handful of observations carefully selected from the tens of thousands that are available in Domesday Book. By visually inspecting the relationship between manorial size and income on the one hand, and the unit of assessment (hides) on the other for only twelve villages, Round concluded that 'Assessment bore no ratio to area or to value in a vill, and still less in a manor. Assessment was not objective, but subjective'.[18] Assessment, according to this interpretation, resulted from the casting down on the vills of England, irrespective of their economic capacity, artificial assessments defined in multiples of five hides. This anti-rational interpretation was subsequently adopted by other leading historians – such as F.W. Maitland, P. Vinogradoff, and F.M. Stenton – and became the traditional view of the behaviour of the medieval state. Even the most influential modern scholar of Domesday Book, Sir Clifford Darby (who published five detailed volumes between 1952 and 1967 and a summary volume in 1977, which was reprinted in 1979 and 1986),[19] concluded on numerous occasions that 'The assessment was largely artificial in character, and bore no constant relation to the agricultural resources of a vill'.[20] More recently, this argument has been generalised by E. L. Jones to help explain the alleged absence of sustained growth of per capita income in pre-modern economies:[21]

> The difficulties of raising average income because of initially weak technologies, feeble institutions, and environments of high natural risk were real. It is not the intention to discount them. There was added however a more central disability: political power based on unchecked violence or the unconstrained threat of violence, which raised the social risk facing the entrepreneur. High and *arbitrary* taxation, even confiscation, was a powerful disincentive and the probability of penalties for conspicuous success even more so.

A new interpretation

Despite the longevity of the traditional interpretation there are good *a priori* reasons for doubting its validity. In view of the inevitable opposition to an arbitrary tax, it is hard to imagine such a system surviving for almost 200 years, as did the geld. Even if suppression of the inevitable riots had been effective, the costs involved would have drastically reduced the net revenue to be gained from imposing the tax. And in turn this would have endangered the survival of the Conqueror's realm in England and in France because of the constant threat of invasion from hostile neighbours.[22] It is this inconsistency that generated my interest in the issue. Clearly, the question is an important one: was the king attempting to impose his authority arbitrarily upon his subjects, as the traditional interpretation suggests, or was he acting in a rational economic way by attempting to maximise his net tax revenue? A definitive answer to this question would provide a major insight into the behaviour of the medieval state.

Since undertaking the initial work on this issue, the fascinating case of Margaret Thatcher's poll tax (or Community Charge) has conveniently emerged. The tax, which was a charge on people rather than on the rateable value of property, was introduced into Scotland in April 1989 and England and Wales in April 1990. Despite the political arguments advanced in its support (to curb the excessive rates of extravagant Labour Councils), this is an excellent example of an assessment system that is arbitrary and economically irrational because it was *not* based upon the equitable principle of capacity to pay. It is precisely what I had in mind in 1985 when I argued that had the geld been an arbitrary tax it would have self-destructed owing to the widespread opposition that would have been generated. Needless to say, I followed the fortunes of the Thatcher poll tax with considerable interest. Predictably, this tax, even in amended form, became so unpopular owing to its perceived lack of equity, that the Conservative Party's ratings dropped rapidly to the point where Mrs Thatcher was forced to resign in November 1990, just seven months after the tax was imposed upon England and Wales. In March 1991 the poll tax in its original form was abolished. And it remains to be seen whether the Conservatives can repair this loss of popularity in time to avoid defeat at the next general election. Such are the dramatic and predictable consequences of the use of arbitrary power. It is for this reason that successful leaders, such as the late Anglo-Saxon and Norman kings, pursued economically rational policies.

The Domesday Survey of 1086, despite Clapham's denial reported in Chapter 2, provides the abundant detail necessary to test the relationship between tax assessments and the capacity of manors to pay this tax, measured in terms of both their resource endowments and their revenues in a feudal European economy. As can be seen from a sample Domesday Book entry (see Figure 5.1 on page 172), data are available on a manorial basis for the following variables: the unit of geld assessment (numbers of 'hides'); the annual 'value' (*valet*) or income; the number of ploughteams; the number of free and dependent labourers; the amount of woodland, pasture, and meadow; numbers of mills, saltpans, fisheries, vineyards and so on; and, for three of the thirty-one counties, numbers of cows, sheep, swine, goats, and horses. To reduce the number of explanatory variables to a manageable size, some – namely the components of labour and livestock – have been combined, and others – saltpans, fisheries, and vineyards – have been excluded because they appeared incomplete or were minor resources. As the capacity of manors to pay tax can be measured by either income or resources, two sets of explanatory variables have been employed in the regression analysis: income (*V*), on the one hand, and the resource variables of ploughteams (*Pl*), livestock (*Li*), labour (*N*), slaves (*S*), woodland (*W*), meadow (*Me*), pasture (*P*), and mills (*M*), on the other.

Before discussing the results of the regression analysis, two statistical problems that are important in the case of Domesday Book data need to be considered. First, while tax assessment was expected to be an increasing function of manorial income or resources, theory does not indicate the precise functional form of the relationship. Rather than just assume that it was linear or log-linear, the Domesday data were employed to determine the choice of the functional form. This was done by assuming that the appropriate function is a member of a broad class of functional forms as defined by the Box–Cox extended model, in which the linear and log-linear forms are special cases. As it turned out, this was a sensible precaution because the linear form was not appropriate in any of the economic relationships so far examined in this project and the log-linear form was only a reasonable approximation for one economic relationship (that between assessment and values). Secondly, the regression analysis of Domesday data has to cope with heteroskedasticity (where the variance of the disturbance is not constant over the sample, but is related to the size of the unit of measurement), a common problem when using cross-sectional data. To overcome this problem, the estimation method used here allows for the disturbances to be heteroskedastic, with the variance of the disturbance depending upon the size of the manor.

The regression work, which has been undertaken for a large number of counties involving thousands of observations, has generated remarkably consistent results for both lay and ecclesiastical manors. Accordingly, only two tables – Tables 6.1 and 6.2 – are presented in this chapter. Also, as these results have been well documented in a series of publications since the mid-1980s, only a few general observations will be made here. Table 6.1 provides the regression results for the tax assessment/capacity to pay relationship for lay manors in Essex and Wiltshire. The first point to note is that, for both counties, all the regression equations, which pass tests of the critical assumptions underlying regression analysis,[23] provide remarkably good fits ($\bar{R}^2$ ranging from 0.63 to 0.79 for the relevant equations (1) and (3)) for disaggregated cross-sectional data. Secondly, in Equation (1), which tests the assessment–resources relationship, the main resource variables of ploughlands, peasant labour, slaves, and livestock (available for Essex only), are both positive and significant. Thirdly, the assessment–income relationship (regression 3) is also positive and highly significant, and the elasticity estimate suggests that, on average, a 1 per cent increase in manorial income is associated with a 0.71 per cent increase in tax assessment. The slightly better fit for the income relationship appears to suggest that the geld was an income rather than a resources tax. Table 6.2 provides further and similar results at an aggregated level for twenty-nine Domesday counties.

What do these results tell us about the nature of the geld? The graphical reconstruction in Figure 6.1 suggests that the geld was a regressive tax levied on the income of manors and that it could have resulted from striking a few basic marginal assessment rates. The evidence in Domesday Book also indicates that certain favoured individuals received partial or complete relief from this tax impost (known as 'beneficial hidation') in recognition of their power and usefulness to the monarchy. Therefore, by carefully measuring manorial income (and resources) and basing this taxation system on the principle of capacity to pay, subject to the political constraint of placating his powerful barons, the king appears to have attempted to maximise his net tax revenue. This policy could have been implemented quite easily by the king's agents had they been supplied with a simple scale of marginal assessment rates expressed in numbers of hides, together with a list of barons to whom concessions were to be granted. Needless to say, this is hardly the action of an absolute king wilfully exercising his arbitrary power, but rather the policy of a powerful but prudent monarch – more prudent than his political heir some 900 years later – grappling with the political realities of his time in an attempt to maximise his net tax revenue so as to achieve his territorial ambitions.

Since first presenting this rationality interpretation,[24] there has been an unsuccessful attempt to reassert the artificial assessments argument, by claiming that econometric techniques are not appropriate to a study of Domesday data and by turning once again to the evidence that, for *some* counties, hides appear to be arranged in multiples of five at the vill level.[25] This reactionary assertion ignores the overwhelming evidence of the Domesday data that assessments were based upon capacity to pay and had nothing to do with any arbitrary five-hide rule of thumb. As shown here, the five-hide and the capacity-to-pay

Table 6.1 Tax assessment–capacity to pay relationships for Essex and Wiltshire lay manors, 1086

		Constant	*Pl*	*Li*	*N*	*S*	*W*	*Me*	*P*	*M*	*V*	$\bar{R}^2$
I Essex												
Regression 1	β	8.11	1.17	0.051	0.391	0.218	0.035	−0.206	0.111	0.048		0.626
	t	32.3	9.6	4.2	5.1	3.0	1.5	−4.6	4.5	0.3		
	E		0.43	0.06	0.18	0.07	0.03	−0.10	0.06	0.01		
Regression 2	β	8.35	1.94									0.552
	t	71.9	28.9									
	E		0.72									
Regression 3	β	2.41									0.709	0.642
	t	29.0									34.9	
	E										0.71	
IIWiltshire												
Regression 1	β	14.07	2.12		0.734	0.630	−0.009	0.478	0.132	−0.366		0.737
	t	30.3	8.3		4.8	3.7	−0.3	6.1	4.4	−1.2		
	E		0.36		0.18	0.09	−0.01	0.13	0.07	−0.04		
Regression 2	β	15.08	3.75									0.658
	t	73.0	28.5									
	E		0.71									
Regression 3	β	2.66									0.808	0.794
	t	33.7									40.5	
	E										0.81	

Notes: The Essex regressions were based on 678 observations and the Wiltshire regressions on 425 observations. The Essex and Wiltshire regressions 1 and 2 are Box–Cox extended transformations, with $\lambda = 0.230$ for the Essex regressions; $\lambda = 0.340$ for Wiltshire regression 1; and $\lambda = 0.320$ for Wiltshire regression 2. The Essex and Wiltshire regressions 3 are of log-linear form.

The first row (labelled β) gives the parameter estimates of the intercept or constant and the following explanatory variables: ploughteams plus excess land for additional ploughs (*Pl*), livestock (*Li*), peasants (*N*), slaves (*S*), woodland (*W*), meadow (*Me*), pasture (*P*), and mills (*M*). (*V* refers to annual value.) The second row (labelled *t*) gives the conventional *t*-ratios, and the third row gives the partial resource–tax assessment elasticities evaluated at the sample mean of tax assessment and the resource. For Wiltshire regressions 1 and 2, *Pl* refers to ploughteams only.

The Essex and Wiltshire regressions 3 appeared appropriate when a functional form test was performed, and for all regressions the hetero (*a*) and hetero (*b*) tests did not reveal significant heteroskedasticity in the disturbances.

Source: Snooks and McDonald, *Domesday Economy*, p. 63.

Table 6.2 The tax assessment–income relationship for twenty-nine counties in Domesday England, 1086

		Constant	*Income (value)*	$\bar{R}^2$	*s*	*F-Form*	*Hetero (a)*	*Hetero (b)*
Box–Cox extended model	β	4.50	0.774	0.719	3.70		1.59	2.50
$\lambda = 0.290$	t	1.7	8.5				t(24)	F(2.23)
							±2.06	3.42
	β	0.26	0.938	0.814	0.491	2.18	−0.53	0.59
Log-linear model	t	0.4	11.1			t(25)	t(24)	F(2.23)
							±2.06	3.42

Notes: The following counties were excluded from the regressions because of the devastation they suffered at the hands of the Conqueror: Cheshire, Derbyshire, Nottinghamshire, Shropshire, Staffordshire and Yorkshire. The rows labelled β give the coefficient estimates and the rows labelled t give t-ratios. $\bar{R}^2$ is the coefficient of determination adjusted for degrees of freedom and s is the disturbance estimated standard error. F-Form, hetero (*a*) and hetero (*b*) relate to functional form and heteroskedasticity tests discussed in Snooks and McDonald, 'Tax assessments', *Economic History Review*, 1985. In the first row the test statistic is given and underneath the distribution of the test statistic under the null hypothesis and the 5 per cent critical values.

Source: Snooks and McDonald, 'Cliometric analysis', *Economic History Review*, 1987, p. 260.

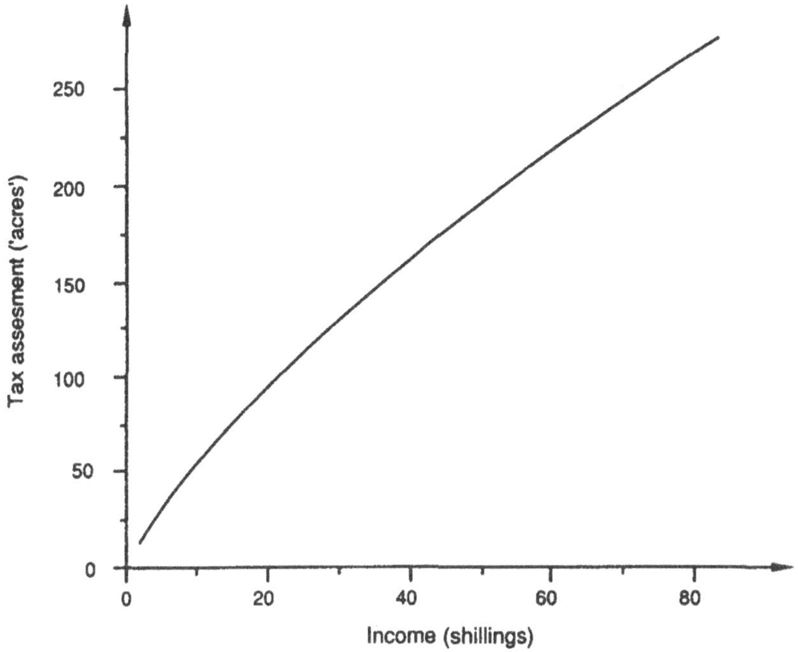

Figure 6.1 The relationship between tax assessments and manorial income, Essex lay manors, 1086

Source: Snooks and McDonald, *Domesday Economy*, p. 65.

hypotheses are mutually exclusive explanations. If one is correct, as the capacity-to-pay hypothesis was found to be, then the other is incorrect: a logical imperative that appears lost on some critics. Although there is overwhelming evidence to show that the five-hide unit is irrelevant to an explanation of the assessment system, I thought it might help to lay this issue to rest by providing an alternative explanation for the occurrence of the five-hide unit in *some* counties. In a recent paper I argued that in some counties the hidage system appears to have been employed to define local authority boundaries at the level of vill and hundred.[26] Evidence was presented to show that the vill, like the hundred, was an administrative unit based upon the amount of land in its constituent manors, and that the size of a vill (or hundred) could be, and was, adjusted for administrative convenience. Hence while the five-hide unit was not used as a rule of thumb to determine the distribution of tax assessments it was employed in *some* counties to determine local authority boundaries.

The economic role of manorial lords

The traditional interpretation

One of the central issues to emerge from an analysis of data in Domesday Book is the nature of the relationship between manorial income and resources. Its importance derives from the information provided about the nature of manorial entrepreneurship in feudal England. Currently there is a vaguely-articulated view held by many medieval historians, stretching back at least to Cunningham, that economic man is a modern development. According to this view, medieval man was motivated largely by considerations of what was in the community's best interest – of what was just – and which were enshrined in fixed rules of conduct, generally referred to as custom. In the field of Domesday studies this view is manifested in the refusal to see any systematic relationship between the income (or values) and the resources of manors.

The view that the values were 'artificial' can be traced back at least to the late nineteenth century. F. W. Maitland, the great historian of his age, thought that for many Domesday manors 'there are cases in which the *valuits* and *valets* look as artificial and systematic as the hides and the teamlands . . . Everywhere we are baffled by the make-believe of ancient finance'.[27] This representative attitude of Victorian scholarship has had considerable influence up to the present time. Darby, for example, not only accepted the Victorian attitude to assessments but also, like Maitland, extrapolated that attitude to the manorial values. Throughout his various publications on the geography of Domesday Book, Darby claimed repeatedly, after formally examining only a handful of observations, that 'it is impossible to discern any consistent relationship . . . between resources and values'; that 'It is clear that the values were not directly related to the number of working teams or to the recorded population. Nor does a consideration of other resources help us to understand the figures'; and that 'No constant relationship between resources and values appears'.[28]

A new interpretation

The implication of the traditional view about manorial values is that the Normans were either unable to determine and clearly document realistic estimates of income, or that they lacked either the desire or the ability to manage their estates effectively, or a combination of both.

None of these possibilities fits the usual picture of the Normans as efficient administrators in what has been described as 'a community organized for war'.[29] In the first place, it makes little sense to conclude that an administration able to conduct such an impressive survey in such a short time would collect meaningless data on income (yet meaningful data on the other main variables that curiously have the confidence of medievalists), or compile it in such a way that the meaning could not be reconstructed. Secondly, if the Normans possessed the desire and ability to be effective administrators of state and army, why should they lack these qualities when it came to managing their own estates, particularly as the latter was essential to the former? Finally, in a period when the efficient use of agricultural resources was required to prevent both invasion and starvation, how could medieval man afford to be anything other than economic man? These fundamental doubts fuelled the determination to investigate the income-resources relationship further.

Multiple regression analysis was employed to test the association between manorial income and resources. With income (*V*) as the dependent variable, the following ten explanatory variables for each manor were constructed: ploughteams (*Pl*), livestock (*Ll*), horses (*H*), labour (*N*), slaves (*S*), woodland (*W*), meadow (*Me*), pasture (*P*), mills (*M*), and beehives (*B*). Once again the data were employed (through use of the Box–Cox model) to determine the functional form of the relationship, because economic theory merely tells us that income increases as more resources are employed; and heteroskedasticity was allowed for in the regression equation. The relationship was tested by using thousands of observations from a large number of counties. Representative results, which pass all the diagnostic tests, for Essex and Wiltshire lay manors are presented in Table 6.3. In regression 1, all the main resources (except mills and beehives)[30] are positive and highly significant and the estimated functions provide unusually good fits ($\bar{R}^2$ of up to 0.87 for lay manors and 0.92 for ecclesiastical manors) for disaggregated cross-section data. Notice also in regression 3 that the horse variable is insignificant, which confirms the suggestion in the literature that horses were used only for war, leisure, and public display – oxen were the beasts of agriculture.

Contrary to the conventional wisdom, therefore, these results demonstrate a very strong and positive relationship between manorial income and resources. The implications are clear. First, any lingering doubts about the reliability of Domesday Book data, particularly the much maligned values, must be dispelled once and for all. The

Table 6.3 The annual value–resources relationship for Essex and Wiltshire lay manors, 1086

		Constant	*Pl*	*Li*	*N*	*S*	*W*	*Me*	*P*	*B*	*M*	*H*	$\bar{R}^2$
I Essex													
Regression 1	β	4.52	1.35	0.042	0.307	0.311	0.034	0.113	0.055	0.0007	−0.111		0.865
	t	26.2	17.9	6.4	6.7	6.7	2.8	4.2	4.1	0.1	−1.2		
	E		0.55	0.07	0.17	0.11	0.03	0.06	0.04	0.00	−0.02		
Regression 2	β	4.67	1.34	0.041	0.303	0.308	0.033	0.106	0.055	0.005			0.865
	t	39.4	17.8	6.3	6.7	6.7	2.7	4.1	4.1	0.09			
	E		0.55	0.07	0.17	0.11	0.03	0.06	0.04	0.00			
Regression 3	β	4.48	1.35	0.044	0.307	0.312	0.034	0.112	0.055	0.007	−0.104	−0.039	0.865
	t	24.4	17.8	5.5	6.7	6.7	2.7	4.2	4.1	0.1	−1.1	−0.6	
	E		0.55	0.07	0.17	0.11	0.03	0.06	0.04	0.00	−0.02	−0.01	
Regression 4	β	4.86	1.43		0.343	0.350	0.037	0.101	0.065				0.856
	t	53.5	18.7		7.4	7.4	2.9	3.8	4.7				
	E		0.58		0.19	0.12	0.04	0.06	0.05				
II Wiltshire													
Regression 1	β	5.10	0.868		0.406	0.260	−0.014	0.101	0.120		0.048		0.803
	t	30.8	9.5		6.8	4.2	−1.0	3.2	8.3		0.5		
	E		0.36		0.22	0.10	−0.01	0.06	0.12		0.01		
Regression 2	β	5.04	0.870		0.411	0.260	−0.013	0.103	0.120				0.804
	t	46.5	9.6		7.0	4.2	−0.9	3.3	8.4				
	E		0.36		0.23	0.10	−0.01	0.06	0.12				
Regression 3	β	5.52	0.858		0.435	0.264	−0.017	0.101	0.123		0.285		0.814
	t	31.8	9.0		7.0	4.1	−1.2	3.1	8.3		2.7		
	E		0.34		0.23	0.10	−0.02	0.06	0.12		0.07		

Notes: The dependent variable in all regressions is the annual value, except in Wiltshire regression 3, where it is the sum of annual value and mill render. All relationships are Box–Cox extended transformations (for the Essex regressions, which were based on 682 observations, $\lambda = 0.280$; for Wiltshire regressions 1 and 2, based on 425 observations, $\lambda = 0.260$; and for Wiltshire regression 3, based on 415 observations, $\lambda = 0.270$).
The first row (labelled β) gives the Box–Cox extended parameter estimates of the intercept or constant and the following explanatory variables: ploughteams (*Pl*), livestock (*Li*), peasants (*N*), slaves (*S*), woodland (*W*), meadow (*Me*), pasture (*P*), bees (*B*), and mills (*M*). (*H* refers to horses.) The second row (labelled t) gives the conventional t-ratios, and the third row gives the partial resource–annual value elasticities, evaluated at the sample mean of annual value and the resource. For all regressions, the hetero (*a*) and hetero (*b*) tests did not reveal significant heteroskedasticity in the disturbances.

Sources: Snooks and McDonald, *Domesday Economy*, p. 87; Snooks, 'Arbitrary decree or rational calculation', *Australian Economic History Review*, p. 39.

remarkably close fits between the estimated functions and the data – fits rarely found using modern cross-sectional data – are a measure of the unusual accuracy of this source discussed elsewhere.[31] The Conqueror's thorough and costly survey methods are surely vindicated by the enviably high quality of the data collected. Secondly, these results demonstrate clearly that the economic responses of manorial lords were remarkably consistent despite the regional diversity in geography, climate, and political and cultural heritage throughout England. The most convincing explanation of such high goodness-of-fit values is that manorial lords were attempting to maximise their incomes. Any other hypothesis, such as the traditional one of manorial lords responding to 'conventional rules' or custom, is unsatisfactory. If the 'conventional rules' concept is to have any meaning, these rules would have to be resistant to changing economic circumstances across regions and across time – otherwise they would be no more than a fiction, masking the attempt by manorial lords to respond to changing circumstances. If, however, the production rules were inflexible over the longer term in the face of regional economic differences, and if they were, in fact, rigidly adopted by manorial lords, then we would not obtain such high goodness-of-fit coefficients. There is little doubt that the production 'rule' generally adopted by manorial lords was that of maximising behaviour. Once again we are led to the conclusion by overwhelming evidence that medieval man was not moral man as the traditional view would have us believe, but rather economic man. Over the last millennium, therefore, the structure of human society may have changed radically, but fundamental human nature has remained the same.

The manorial production process

The central microeconomic institution in the feudal economy of Domesday England was the manor which, unlike the modern firm, was the basic unit of consumption as well as production. While much has been written about the theory and history of the manor it is remarkable how little is known about the processes underlying the manorial economy.[32] Little is known about the technical relationships between manorial inputs and outputs, about the conditions of factor substitution, or about the nature of returns to scale. This may not seem surprising in view of the antiquity of the economy in question, which flourished nine centuries ago. But surprising it is, because the data required to analyse the production relationships in late-eleventh-century England have been readily available to scholars for the last few hundred years in various facsimile editions of Domesday Book; and because these data are far more comprehensive and reliable than comparable data today. All that we are told about these matters by the traditional scholars is that the manorial inputs, such as oxen and labourers, were used in fixed relationships and that they were '*in no way interchangeable in practice*'.[33] In other words, the elasticity of substitution of factors was zero! This interpretation is part of a wider, although not clearly articulated, view that early medieval agriculture was a largely static and inflexible system that did not allow for much individual initiative or adaption to variations in local conditions. It was, we are told, the prisoner of custom.

A new approach to the feudal system of manorial production

By the use of production function analysis, it can be shown that the system of manorial production was both very flexible and capable of supporting considerable individual initiative. To interpret the Domesday production function estimates at face value we need to view the manorial system as one in which the landholder maximises manorial income, subject to feudal institutional constraints – mainly of undeveloped factor markets – that lead to fixed input levels in the shortrun. While this model is based upon the descriptive detail given in Chapter 5, it does not pretend to encompass all aspects of that historical reality; rather, it is an abstract characterisation of the way in which the manorial economic system worked. This model provides a more satisfactory basis for interpreting the Domesday production results than that usually employed by economists, in which a profit-maximising firm in a perfectly competitive environment is able to vary its input levels in the shortrun. The extent to which this characterisation is not true is a measure of the care we should take in accepting the estimates at face value. It is thought, however, to be a reasonable approximation particularly in view of the evidence presented here and in Chapter 5.

A virtue of this model is that it can explain why manors were operated with widely different resource ratios (such as the ratio of peasants to ploughteams) and at very different production levels. The remarkable variation in manorial size and resource combination recorded in Domesday Book can be accounted for by the history and geography of a particular region, natural processes such as fertility and disease, and the requirements of a feudal society. A lord being granted a manor by the Conqueror received a set of endowments, which included a precisely defined amount of land together with a given number of peasants, slaves, ploughteams, livestock, and other assets of various types. In the short term there was little the new lord could do to change either the maximum scale of production or the combination of resources because, although there was a market for output, as seen in Chapter 5, there was little trading in capital and labour resources. The resources could be increased only by long-term breeding programmes that could be suddenly and adversely affected by unexpected droughts, floods, disease, and Viking raids. In these circumstances, the only short-term option available to the manorial lord was to take the amount of each resource as given and attempt to organise production as efficiently as possible in order to maximise the net value of the goods produced.

A Domesday production function

The Domesday economy is an ideal testing ground for modern production theory. Because of the relatively simple nature of this economy, we do not have to contend with a number of serious problems that normally bedevil production functions estimated for modern economies, namely simultaneity (as inputs are exogenously determined) and the conceptual and measurement problems concerning capital stock (as the basic item of capital was the number of ploughteams). Also a simple economy, such as that which existed in Domesday England, provides a less ambiguous test for economic theory than a highly sophisticated and complex modern economy. As will be demonstrated, while growth theory has virtually no application to historical growth processes, static neoclassical theory comes into its own, but only when the clock of history is stopped.

In the literature a variety of production functions has been applied to data for modern firms. Often the function chosen is not the most appropriate one, but rather that which is easiest to estimate. Ease of estimation, however, does not necessarily make for accuracy of interpretation. Unfortunately, the simple production functions make strong assumptions about the nature of the production process which, in most cases, force the data into a very restricted form and limit what they can tell us about the production system. The most popular production function is the Cobb–Douglas form which, although intuitively appealing, makes a number of very strong assumptions.

To enable ease of estimation, this model imposes a predetermined log-linear functional form upon the data, rather than allowing the data to select the appropriate form. In turn, this imposes strong assumptions on the substitution of factors, such that villeins can only be substituted for ploughteams in a particular manner or, to put it another way, that the isoquant has a predetermined shape. The reason is that the elasticity of substitution for the Cobb–Douglas function is set equal to unity. Even the CES function, while allowing different substitution properties, assumes that they are constant for all combinations of ploughteams and villeins (or at all points on the isoquant). Also, when more than two inputs are taken into account, the substitution behaviour between all pairs of resources – say, villeins and ploughteams, ploughteams and livestock, villeins and slaves – is assumed to be the same.

Ideally, we wish to employ a production function that allows the data to be used to estimate all properties of both substitution and return to scale. This is possible with a very flexible model called the Generalized Quadratic function,[34] but the computing difficulties are formidable because in our case it is necessary to estimate over fifty parameters. As a compromise, more flexible forms of the CES function, including the Sato Two-Level form, have been employed.[35] In the Sato function, the inputs are grouped into the three 'higher' level categories of land (pasture, meadow, and woodland), labour (freemen–sokemen, villeins, bordars, and slaves) and capital (demesne ploughteams, peasant ploughteams, and livestock). Pairs of resources have the same substitution behaviour if they come from the same 'higher' level category, but the substitution behaviour can differ if they come from different 'higher' level categories. The Domesday data were used to estimate the appropriate substitution behaviour at each level.[36]

A reconstruction of the microeconomic system of Domesday production

Comprehensive statistical analysis of the system of manorial production is possible only for Essex. Of all the counties, only Essex has detailed and unambiguous data for all the resources employed upon Domesday manors, and this only for 1086.[37] The main results presented in Table 6.4, regression 1, are generated by a CES production function for which the conditions of substitution and returns to scale are estimated using disaggregated data. These estimates show that all coefficients (with the exception of beehives) are positive and significant, and the coefficient of determination ($\bar{R}^2$) is a remarkably high 0.969. The CES production function therefore provides an excellent fit for the Domesday data on net production and economic resources.[38] As this function assumes that all pairs of inputs experience the same resource substitution conditions, further results are given – Table 6.4, regression 2 – from the estimation of a more flexible (the Sato Two-Level) production function that allows variation in this matter. As the characteristics of the Sato function are broadly similar to the CES results the story about manorial production can be based upon the results in regressions 1 and 2.

The results in Table 6.4 show[39] that manorial production increased as resources or inputs increased, a conclusion formerly denied by a leading Domesday scholar – Sir Clifford Darby;[40] that diminishing marginal rates of substitution were experienced; that there was, in contrast to the conclusion of traditional scholars, a high degree of substitutability of resources – the elasticity of substitution was 1.76, which is clearly not zero as suggested by traditional scholars(!); and that modest increasing returns to scale (the combined partial output elasticities amounted to 1.15) prevailed. And modest increasing returns are consistent with the positive rates of growth measured in the following chapter. These results also confirm the dominance of arable farming, which was 2.6 times as important as pastoral activities; suggest that demesne ploughteams made a greater contribution to production than peasants' ploughteams (by a factor of 2.6); and also suggest that the role of villeins was greater than that of bordars (partial elasticities were 0.15 and 0.11 respectively), which in turn was greater than that of slaves (with a partial elasticity of 0.08) – indeed, slaves appear to have worked to a greater degree in the lord's household than has hitherto been suspected. It can be concluded, therefore, that the application of relatively sophisticated production functions to the data in Domesday Book provides significant insights into the process of production in the English feudal economy. Previous scholarship, which is hampered by an absence of information other than that in Domesday Book (which it has been unable to harness adequately) has described the type of economic institutions in this period, but has been unable to show exactly how the production process worked. Indeed they have claimed that manorial production was the prisoner of custom, whereas a systematic examination of the Domesday data shows that it was a flexible and competitive system that responded to market signals. These results also demonstrate the usefulness of modern production theory in making it possible to reconstruct a feudal microeconomic system that flourished over 900 years ago – a system

Table 6.4 CES production functions for Essex lay manors, 1086: non-linear least squares estimates using disaggregated input data

	λ	μ	β^*_0	β^*_1	β^*_2	β^*_3	β^*_4	β^*_5	β^*_6	β^*_7	β^*_8	β^*_9	β^*_{10}	β^*_{11}	$\bar{R}^2$	F
	Explanatory variable			*Pld*	*Plp*	*Li*	*FS*	*V*	*Bo*	*S*	*W*	*Me*	*P*	*B*		
Regression 1: CES																
Coefficient estimate	0.432	1.430	0.79	0.895	0.333	0.018	0.087	0.244	0.134	0.177	0.014	0.059	0.032	−0.042	0.969	1648.1
t-ratio	10.7	15.5	5.1	6.2	4.0	4.8	2.1	4.6	4.2	3.9	2.4	3.4	4.0	−1.2		*F*(13 669)
Elasticity				0.41	0.16	0.09	0.03	0.15	0.11	0.08	0.03	0.06	0.04	−0.01	$\sum=1.15$	2.15
Regression 2: Sato																
Coefficient estimate	0.320	1.265	0.38	1.207	0.746	0.0016	0.0029	0.0704	0.0029	0.0774	0.00003	0.0007	0.00016		0.972	1943.2
t-ratio	7.6	12.6	2.1	2.0	1.8	2.0	0.4	1.4	1.4	1.5	0.7	0.9	0.9			*F*(12 670)
Elasticity				0.32	0.23	0.11	0.00	0.15	0.10	0.09	0.02	0.06	0.03		$\sum=1.11$	2.22
	λ	μ	β_0	α_1	α_2	α_3									$\bar{R}^2$	F
Regression 3: GQ															0.972	4721.8
Coefficient estimate	0.311	1.341	0.47	−0.249	−0.227	0.136										*F*(5 667)
t-ratio	7.7	14.5	3.6	−1.0	−0.9	0.3										3.05

Notes: *Pld* is a variable measuring the number of demesne ploughteams on the manor; *Plp* the number of peasants' ploughteams; *Li* the number of livestock; *FS* the number of freemen and sokemen; *V* the number of villeins; *Bo* the number of bordars; *S* the number of slaves; *W* the amount of woodland; *Me* the amount of meadow; *P* the amount of pasture; and *B* the number of beehives. The elasticities are partial output elasticities evaluated at the sample mean of the resources.

$\sum$ is the sum of the output elasticities, that is, the elasticity of scale evaluated at the resource means. $\bar{R}^2$ is the coefficient of determination allowing for degrees of freedom; and *F* is the *F* test statistic for the null hypothesis that $\beta^*_i=0, i=1,2\ldots11$. Underneath are given the degrees of freedom under the null, and the 1 per cent critical value. For all equations, the hetero (*a*) and hetero (*b*) tests did not reveal significant heteroskedasticity in the disturbances.

Source: Snooks and McDonald, *Domesday Economy*, p. 223.

that was far more flexible and rational than the traditional interpretation would have us believe.

A TIME FOR ECONOMIC MAN

The results of this statistical analysis of Domesday Book data suggest that, at the beginning of the last millennium, the primary motivation of decision-makers at both the macro and micro levels of the economy was material self interest. On average, only those non-economic ideas and ideals that were consistent with this goal were pursued on a systematic, widespread, and persistent basis. The Domesday data are just not consistent with the traditional view, both historical and economic, that medieval man was moral man. Hence exactly one hundred years after their debate in the *Economic Journal* for 1892, we can conclude that Marshall rather than Cunningham was right about the role of economic man in the distant past. Medieval man and modern man therefore have much in common, suggesting that human motivation throughout time is basically unchanging.

We have also seen how order and chaos were manipulated by the ruling elite to maximise their incomes. In the input market the tenants-in-chief (1000 in number) and their under-tenants (a further 6100 men) combined to become monopsonists so as to extract as much surplus as possible from the peasantry. In the commodity market, which the elite did not control because possibly 75 per cent of it was exported, they competed vigorously with each other. And in the political arena the competition between barons was also fierce and was conducted through economic rather than military channels owing to the ban the Conqueror had placed on private warfare in England. But as feuding barons could meet in combat in their home country of Normandy, they would have had an additional incentive to maximise their income from their English estates. Quite simply, the greater the income the more mercenaries they could employ in France. And those who won the political/military battle could also expect to receive further economic favours from the king. There can be little doubt that it was a time for economic man.

So far we have seen that market forces had made considerable inroads into the economic system of feudal England. At least 40 per cent of national income was generated in the market economy. Further, we have discovered that economic agents in feudal society were sensitive to market signals, in that they appeared to be acting in economically rational ways. What remains is to determine whether the ruling elite in England was sufficiently successful in its pursuit of economic goals to achieve rapid and sustained growth.

7 Accounting for the Very Longrun

Long is the way and hard

(John Milton, 1667)

IS GROWTH A MODERN INVENTION?

A widely-held view amongst historians, economists, and ecologists is that economic growth – defined in terms of a sustained increase in GDP per capita – is a modern invention. While there are exceptions, the vast majority of scholars, both in the social and natural sciences, take it as read that rapid growth began with the Industrial Revolution, and that any growth before 1700 was very slow, and before 1500 was non-existent.[1] The conventional wisdom tells us that economic systems in the distant past experienced, over very long periods of time, either the steady state envisaged by the classical economists, or zero-sum fluctuations in GDP per capita. According to both interpretations, ancient and medieval societies were unable to escape from poverty because they were dominated by custom rather than individual self interest. The implications of such a conclusion are fundamental to our view of the past and future of the human race. Ecologists, for example, believe that just as growth was turned on at the beginning of the Industrial Revolution, so it can – indeed must – be turned off again immediately.

This view of past societies as static and disembodied is a legacy of both the liberal intellectual conviction that man aspires to higher objectives than mere material self interest, and to the ahistorical approach of economics. The argument presented here is that these interpretations continue to exist only because of the absence of quantitative data required to test them. In the absence of such data, the conviction of historians that medieval man was moral man and not economic man, and the view of classical economists that the dominating landlord class in feudal and medieval society squandered the surplus they extracted from the bulk of the population, remain largely unchallenged.[2] In this chapter an attempt will be made to estimate the change in GDP per capita over the past millennium in order to pose central questions about the nature of feudal and medieval economy, and about the role of growth in human society.[3] It will be shown not only that growth has been a persistent feature of English society over the last 1000 years, but that it has taken place via a number of great waves of economic change.

CAN VERY LONGRUN GROWTH BE MEASURED?

It has long been taken for granted that the quantitative measurement of longrun economic growth in terms of per capita income can be pushed back only to the late seventeenth century – the time of Petty, Davenant, and King. The political arithmetic compiled by these authors constitutes, it is claimed, the very beginning of relatively reliable contemporary estimates of national income. It is worth quoting a number of leading scholars from both the quantitative and non-quantitative traditions in economic history on this issue. P. Deane, one of the pioneers of the British quantitative tradition, claimed in her important evaluation of Gregory King's work that:[4]

> Apart from some early attempts [earlier in the seventeenth century] to estimate the balance of trade of England and to use foreign trade records as indicators of national prosperity, *there are no contemporary assessments of national economic strength of a specifically quantitative nature, until the end of the seventeenth century.*

She goes on to say that owing to the consistent approach adopted by King, together with 'the unprecedented wealth of source material', we can treat 'the end of the seventeenth century as a jumping off point for the study of economic growth in England'.

More recently N. F. R. Crafts concluded that 'estimates of economic growth prior to the mid-nineteenth century for Britain are bound to be in the nature of controlled conjectures rather than definitive evidence'.[5] And here he is referring to the contemporary estimates that began with King – before then even 'controlled conjectures' were not possible. One outcome of the traditional view is that the perception by historical economists of the longrun is defined artificially by the alleged availability of reliable data rather than by historical considerations. Hence the conception of the longrun by historical economists is confined to the period since 1688 – a period of only three hundred years. Once again this attitude is reflected in the recent work of Crafts when he makes the observation that 'The major pioneering work in which estimates of economic growth *over the long run* were first produced was that of Deane and Cole'. Three centuries may be regarded by most economists and many economic historians as a lengthy period, but in fact it merely marks the most recent phase of economic expansion in Europe – the phase of modern economic growth that has taken place within an institutional arrangement which we regard as our own. As we shall see, over the last millennium there have been three such episodes – three great waves – of economic expansion. This restricted approach to the longrun fails to encompass the emergence of modern economic systems from earlier organisational forms and therefore limits our understanding of possible future change.

It is unfortunate, therefore, that the economic historians, generally non-quantitative, who do believe that an analysis of economic change in the very longrun has much to offer our understanding of the present, are forced to deal with the last thousand years in a general and imprecise way. These historians have accepted the judgement of the historical economists that reliable data at the aggregate level does not exist before 1700, or indeed before the 1800s. S. Pollard and D. W. Crossley, for example, claim:[6]

> There is little purpose in trying to measure the wealth of Britain before the late seventeenth century in numerical terms. Before the Political Arithmetic of Petty, Davenant and Gregory King there are no trustworthy contemporary estimates on which to draw, and we have to adopt broader and less precise methods . . . we rely not on figures but on more human evidence.

More recently, E. L. Jones has given his support to this position by claiming:[7] 'Unfortunately, economic statistics from the distant past are very, very poor, or they are simply not available at the macro- or whole society level, so that we cannot go in for real measurement.' Like Pollard and Crossley, Jones also relies upon 'indirect signs of growth', such as life expectancy, technical innovation, structural change, consumption levels, and so on, which are viewed largely in non-quantitative terms. We have, therefore, no overall picture of economic change in the very longrun, with the result that historians have failed to recognise the *systematic* process of growth over the millennia.

Fortunately it is not correct, as the traditional view would have it, that either there are no reliable contemporary estimates of national income for England (at least, before the late seventeenth century) or that it is not possible to measure economic growth with any precision before that time. As will be shown, we can measure economic growth in England over a period of 600 years prior to 1688, with as much precision, if not more, than that for two hundred years after 1688. As shown in Chapter 5, the real starting point for estimating national income, and thereby the rate of economic growth, in England is 1086, the year that Domesday Book was compiled by order of William the Conqueror. There is a role, therefore, for the cliometrician in analysing economic growth in the very longrun. It should also be realised that the data in Domesday Book are more reliable than any other record of national income until the twentieth century.[8] It is difficult, therefore, to understand why economic historians interested in longrun change have ignored the unequalled wealth of detail in Domesday Book, especially as it has a long history of scholarly research. Economic historians from Cunningham to Clapham, to Pollard and Crossley even make indirect use (through Domesday specialists) of selected data from this source, while at the same time denying that any reliable estimates of national income can be made before 1688. It is essential to rekindle this one great statistical beacon of the Middle Ages.

In Western Europe only England possesses a comprehensive economic census dating back as far as the eleventh century. As Cipolla laments: 'We do not have adequate data for measuring in quantitative terms the long-run increase in production and consumption' in Europe.[9] But, like so many others, he had overlooked Domesday Book. Is it possible, therefore, to use England as a rough guide to the nature and rate of economic change over the six hundred years before 1700? There are a number of reasons for answering this question in the affirmative. First, England's relative position in Western Europe was much the same in 1086 as it was in 1688. E. Miller and J. Hatcher quote the Conqueror's chaplain and economic adviser (see Chapter 3), among others from France, as saying that England was a land 'fertile by virtue of its own fecundity, the wealth of which merchants have increased by bringing in riches. Treasures have been amassed there which are remarkable for their number, their quality and their workmanship', and conclude that 'Men like William of Poitiers and Goscelin judged England by the standards of the west in their own time; and by those standards, evidently, it was a richer land, a more advanced land', and that 'by the standards of the time, it [England] was precocious in its capacity for development'.[10] Certainly the various contestants for the English throne from France and Scandinavia in the first half of the eleventh century thought it was worth risking everything they possessed – kingdom and life. Six hundred years later, Gregory King's social accounting suggested that the per capita income of England (£7.90) was marginally less than Holland's (£8.06), but significantly greater than that of France (£6.15). Secondly, England's economic growth did not occur within a vacuum. England was as much an integral part of the Western European economy in the 1080s (when something like a quarter of GDP was exported) as it was in the 1680s. While the timing of economic fluctuations may not always have coincided, the major 'countries' of Western Europe appear to have grown at a similar average rate over the six centuries prior to the Industrial Revolution.

TIME WITHOUT GROWTH? THE CONVENTIONAL WISDOM

As historians claim that there are no national income data for any medieval European country or region before 1700, then they can claim no precision for their best informed guesses about the rate of economic change during the Middle Ages.[11] But because economic growth – the rise from primitive poverty and squalor – is such an important issue, a number of scholars have been tempted to speculate about the rate at which it took place over the longrun in order to provide general explanations of its progress.[12] These scholars fall into a number of categories. The first includes applied economists and historical economists such as Simon Kuznets, Walt Rostow, Angus Maddison, and Gunnar Persson, who use deductive reasoning, and sometimes indirect evidence, to speculate about the rate of growth of per capita income. With the exception of Persson,[13] they appear to believe that economic growth is largely a modern invention, beginning in Europe around 1500 and proceeding very slowly until the 1700s, when it began to accelerate relatively rapidly for the first time. According to this interpretation, stagnation was the norm before the fifteenth century, although some speculate that zero-sum fluctuations probably occurred.[14] The clearest and most careful expression is found in Maddison, *Capitalist Development*:[15]

> For 1500–1700 the rate of progress was also very poor by present standards, but clearly better than in the previous millennium. Per capita output grew at too slow a pace to be perceptible to contemporaries . . . Nevertheless Europe's population grew by a third in these two centuries, per capita output may have risen by a quarter, and total output by around two thirds . . . Productivity rose less than income per head because the increased output required longer working hours.

Maddison, quite sensibly, takes no account of the Brown–Hopkins real wage rate index, which shows for England, France, and Alsace, a substantial and sustained decline from 1500 to 1700![16] For reasons discussed later in this chapter, the Brown–Hopkins nominal wage index must be regarded as highly misleading.

The second category includes historians who are prepared to adopt an intuitive approach, such as John Gould, David Landes, and John Komlos.[17] Gould, for example, is convinced by the apparent logic of the argument that 'it is impossible that the long-run growth rates over the last century should have been sustained for long periods prior to the mid-nineteenth century',[18] but he is also aware of the existence of partial quantitative evidence on economic change in the Middle Ages (unfortunately such as Brown–Hopkins[19]). Sensibly, Gould suggests that this historical evidence is consistent 'with the hypothesis that the period of modern economic growth was preceded by marked secular fluctuations of real income, impressive improvements being cancelled by later and equally impressive declines, *rather* than with the alternative . . . namely . . . centuries-long stagnation or growth at a rate infinitesimally above zero'. But this view is based upon the entirely misleading, as I will demonstrate, real wage index of Brown and Hopkins. More recently, other historians, such as Komlos,[20] who does not refer to the work of Brown and Hopkins or Gould, have re-emphasised in a speculative and non-empirical way the possibility of secular swings within the context of 'thousands of years of very slow economic growth'. Landes begins from a similar position to that of Gould, by claiming:[21]

> We have no true statistical estimates of pre-modern growth; but one has only to extrapolate the levels of income prevalent on the eve of industrialization backward at the rates of growth prevailing after 1700, and one arrives very quickly at levels of income too low for human survival.

He also acknowledges that there were major periods of economic downturn, such as the century or so after the Black Death (but here he does not suggest, possibly wisely, as Maddison does, that there may have been an inverse relationship between population and per capita income) and periodic Malthusian crises. In view of this dismal scenario it is surprising that his guess about the growth of per capita income is more optimistic than Gould's: 'Even so,' Landes concludes, 'it seems clear that over the near-millennium from the year 1000 to the eighteenth century, income per head rose appreciably – perhaps tripled – and that this rise accelerated sharply in the eighteenth century, even before the introduction of the new industrial technology.' Later in this chapter I suggest that while the logic employed by this group of scholars cannot be faulted, their arguments are based upon faulty assumptions.

The final category includes scholars such as Sidney Pollard and David Crossley, Leslie Clarkson, Carlo Cipolla and Eric Jones, who attempt to build up a general picture of changes in wealth and income by examining more indirect, and generally non-quantitative, evidence. Of these only Jones, and only in his more recent work, is willing to draw conclusions about the rate of economic growth. Probably under the influence of Maddison, Jones claims that 'prolonged high rates of growth are not to be found until the Europe of the eighteenth or nineteenth centuries';[22] and that before 1700 growth took place at 'a permanently low rate', which 'asks too much of the usual models. It happens too gradually and over too long a span'.[23] Extending his focus to the whole world over the two or three millennia before 1800, Jones writes: 'The world economy as a whole seems not to have grown in the per capita income sense. On the other hand, population growth and technical change did impart a momentum of continual expansion.'[24] His view, therefore, is that before the Industrial Revolution slow economic expansion (an increase in population with the same level of per capita income) was not uncommon, particularly in Europe, but that economic growth (an increase in per capita income) rarely broke free from the political constraints that confined it to very low positive (or negative) rates. The exceptions that Jones feels confident in nominating are China in the tenth to thirteenth centuries, Japan between 1600 and 1868, and Europe 1500 to 1800 (interestingly, like Maddison, this contradicts the real wage rate indexes of Brown–Hopkins); and even these exceptions have only appeared in his more recent work,[25] possibly under the influence of the historical economists. From his earlier work,[26] one has the impression that he initially believed that significant sustained growth in per capita income was not experienced anywhere in the world, including Europe, before the late sixteenth century. While Jones is to be credited with raising recent interest in the issue of very longrun change, it appears reasonable to conclude that he does not see economic growth as either systematic or rising above zero-sum fluctuations for the pre-modern period.

There is, therefore, little agreement in detail about the possible magnitude of longrun growth rates, which is hardly surprising in the absence of any previous attempt to marshal the national income data required to calculate them. With the possible exceptions of Landes and Persson, however, most of the scholars who have speculated about this matter appear to believe that growth rates of per capita income in Europe were very slow before 1700 and largely non-existent before 1500. Many of these scholars recognise that although growth throughout the millennium before 1700 may not have been significantly different from zero, there were periods of improvement that were largely cancelled out by periods of regression. The exception is Jones who, if I have interpreted him correctly, implies that growth occurred not in terms of systematic economic cycles in individual countries or regions but rather by breaking out almost randomly for political reasons on relatively rare occasions in various parts of the world, possibly never (at least in the pre-1700 period) to occur there again.

The results in this chapter challenge the conventional wisdom that the longrun rate of economic growth in Europe was very slow before 1700, and non-existent before 1500. They also challenge the view that growth was a result not of systematic economic forces, but of political processes, and suggest that the generalised explanations (largely inductive models) that have been developed to explain economic change (or its absence) are highly sensitive to the *assumed* pattern and rate of growth. Instead, it is argued that growth in England – and, by implication, in Europe – was relatively rapid, sustained, and systematic, and that it occurred through the generation of great waves of change that are surging from the past and into the future.

POPULATION GROWTH IN THE VERY LONGRUN

The contours of population growth in England, which are reasonably reliably known, can be employed as a first approximation to an examination of economic expansion during the last millennium. While fairly complete population censuses did not begin until 1801, plausible estimates of population by E. A. Wrigley and R. S. Schofield, based upon demographic data contained within parish records, are available back to 1541.[27] Before this time only approximate estimates are available for the widely scattered individual years of 1525 (based upon tax returns and muster certificates); 1450 (manorial records and indirect evidence); 1377 (poll tax and a sample of manors); 1347 (sample of manors); and 1086 (Domesday Book).[28] Of the medieval estimates, the most reliable is that for 1086, which is based upon the comprehensive Domesday Survey.

From Figure 7.1 it can be seen that the population of England, which increased by a factor of 33.1 from 1.53 million in 1086 to 50.7 million in 1990, experienced three main periods of longrun expansion during this millennium. The first was from Domesday to the Black Death – a period of 261 years; the second from the Tudor revival to Oliver Cromwell – a period of 125 years; and the third, from the Industrial Revolution until the late 1960s – a period of 220 years. Of these, the first and third were quite similar in terms of length – in excess of two centuries – and in terms of the rate of population growth (see the slopes of the population curves in Figure 7.1), although the second occurred slightly more rapidly than the first. Population growth during the sixteenth and seventeenth centuries – the middle period – was quite rapid, but was brought to a close during the 1650s, by which time the pre–Black Death population level had just been re-attained, after a lapse of three centuries. This long downswing in population is the most remarkable feature of Figure 7.1. The second pause in the expansion of population, from 1650 to 1750, is only a shallow and short-lived depression in comparison, as probably will be the levelling off in the last quarter of the twentieth century.

Population, of course, is only part of the story of economic change. We also need to know how the population's average living standards changed over the same period of time. The most coherent view of economic historians on this issue, is that there was an inverse relationship between population and average living standards.[29] This view is based upon the use of the Brown and Hopkins real wage rate index for building workers around Oxford. There are, however, a number of fundamental conceptual problems involved in using this index as a proxy for living standards in England. Even if the index is representative of the entire building industry, it is not necessarily representative of wage rates for all occupations, and even if it were, this would not make it a good proxy for average living standards of the peasant/working class let alone the entire population. This is particularly so in

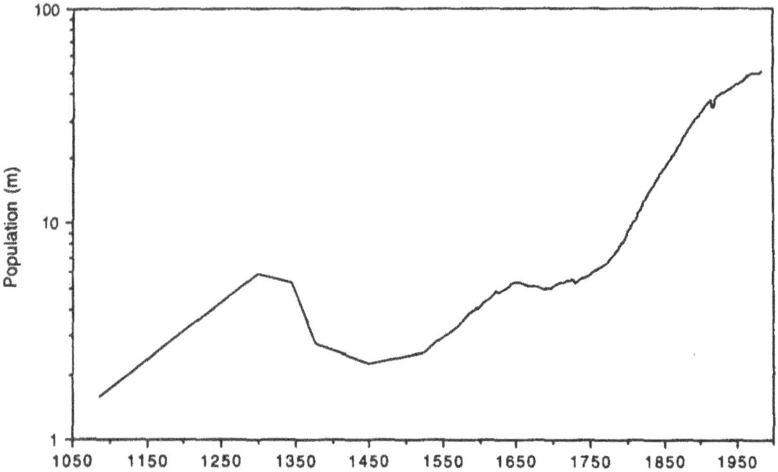

Figure 7.1 The population of England, 1086–1990

Notes: Log scale on vertical axis.
Source: See text.

an economy where the landowning class was able to extract such a large proportion of the total surplus. Both the secular trend and the fluctuations are contradicted by other evidence. And there are good reasons, mainly concerning the nominal wages component, to doubt that the Brown–Hopkins index is a reliable *longrun* measure of even the real wage rates of building workers in Oxford. Accordingly, in the remainder of the chapter an attempt will be made to estimate the growth of GDP per capita over the last millennium.

ESTIMATING MILLENNIAL GROWTH RATES

Domesday Book and Gregory King

A new starting point for the estimation of English growth is provided in Chapter 5. The year 1086 is almost exactly 600 years prior to Gregory King's national income estimates which have for long been wrongly regarded as the earliest relatively reliable macroeconomic data. During the intervening period between Domesday Book and Gregory King, which for the quantitative economic historian is largely shrouded in a mist of obscurity, the economic system of England was transformed from a feudal manorial economy to a mercantile capitalist economy that was on the threshold of a structural transformation later known as the Industrial Revolution. What did this great institutional (or more accurately, systems) change bring with it in terms of material economic growth? We cannot precisely chart the changes from year to year, or even from generation to generation, but we can now evaluate the magnitude of change in material well-being over the entire period of six hundred years.[30]

As we have only two firmly established estimates of GDP per capita spanning six centuries, it is necessary to consider whether these two points in time are strictly comparable. Is there anything unusual about either 1086 or 1688 that will bias my estimate of the rate of growth of per capita income? In the first place, we know that the eleventh century was part of an upswing in economic activity which began in the tenth century and continued until the end of the twelfth century. This expansion was characterised by an increasing colonisation of the unsettled parts of Domesday England – mainly the abundant forests, woodlands, moorland, heath, marshland, fens, and sea marshes – together with a threefold increase in population between 1086 and 1347. Also, there is evidence in the earlier phase of this upswing that economic expansion was also translated into economic growth, but that progressively after the eleventh century the pressure of population growth on increasingly utilised natural resources led to a decline in living standards.[31] Superimposed upon this general trend was the impact of the Conquest of 1066 and the reprisals of the Conqueror against the rebellious north in 1070, which laid waste much of Yorkshire and parts of Cheshire, Derbyshire, Nottinghamshire and Staffordshire. By the time of the Domesday Survey, the English economy was much recovered, although there were still considerable numbers of manors in the north that were 'waste' and abandoned. Yet these abandoned manors were generally located on less productive land, as the surviving population had focused its reconstruction on the more fertile valleys and plains. Hence, in spite of the continuing evidence of devastation, income per household in the north had recovered to pre-Conquest levels. Figure 5.3 (see page 195) shows that even Yorkshire had demesne income levels per household similar to those in adjacent, non-devastated counties. It was the size of the economy, not its average living standards, that had failed to recover to pre-Conquest levels.

England in the time of Gregory King, on the other hand, was growing at a modest rate after fifty years of stagnation following the rapid growth of the sixteenth century, and population was increasing slowly, if hesitantly. This modest expansion and growth was largely a function of the transformation of agriculture during the century after 1660.[32] It would seem, therefore, that 1688 was less favourably placed within its historical context than was 1086. In other words, if a bias has been imparted to my estimates of growth by the 'choice' of beginning and end years, it is one of deflation rather than inflation. In any case the sensitivity analysis in Chapter 5 has shown that even quite large variations in estimates of the level of per capita income have only a relatively small influence upon compound rates of growth calculated over such long periods of time.

Average living standards, 1688 and 1086

An estimate of GDP per capita for England in 1086, which is presented in Table 7.2 (see page 246), has been derived from the income and population estimates given in Chapter 5. The next year for which we have a reasonably reliable estimate of GDP per capita is 1688, calculated by Gregory King. As much excellent work has been undertaken on King's accounts, only a brief discussion is required here. The first point to note is that the estimate by King, whether reworked by P. Deane or P.H. Lindert and J. Williamson, is not as firmly based as that for 1086.[33] Others have adequately detailed the problems with King's work.[34] Nevertheless, when appropriately adjusted, it is widely regarded as providing a relatively reliable basis for measuring the rate of economic progress during the eighteenth century. Even Crafts, despite his words of caution, is prepared to exploit it in his work. Also, the remarkably consistent and plausible econometric results obtained recently by Sir Richard Stone from his estimation of tax and consumption functions using King's estimates suggest that the data are more reliable than some commentators have been prepared to admit.[35] Hence I have accepted King's calculations of national income, population, and per capita income for England and Wales as adjusted by Lindert and Williamson. Secondly, in order to make a comparison of national income between 1086 and 1688 it is necessary to exclude the national income of Wales in the latter year because Domesday Book did not include the subdued but unconquered Welsh tribes. This can be achieved in an approximate way for 1688 by applying the per capita income of England and Wales to the Welsh population and subtracting this subtotal from King's aggregate.

Obviously, this procedure leaves per capita income unchanged. Thirdly, it should be noted that as royal revenues are included in the 1086 national income figure but excluded in 1688, the resulting growth rate calculation will be an underestimate. The results can be found in Table 7.2 (see page 246).

Defining the market and the household

A question of some importance when comparing the national income of 1086 with that of 1688, is the extent to which the market/non-market distinction has been treated in a consistent manner. The objective in this social accounting exercise is to measure that section of total economic activity which modern national income compilers would regard as part of the market sector. For the pre-modern period this means adding an imputed value for subsistence to those activities that were actually measured in the less extensive markets of earlier times. In other words, an attempt has been made to apply a consistent definition of 'market' (including subsistence) and 'household' activities – a definition that places, for example, food growing in the 'market' category, and meal preparation for own-consumption in the 'household' category, as is the modern convention.[36]

Gregory King attempted to measure national income along these lines. He was concerned to estimate the 'income and expense' of families in two categories of roughly equal size – those 'increasing the wealth of the kingdom' (saving), which included twenty-one socioeconomic groups in the middle and upper classes, and those 'decreasing the wealth of the kingdom' (dissaving), which included, in five groups, the employed and unemployed of the working class. In the first category, King estimates the market income from estates, businesses, and salaried occupations; and in the second he estimates the wages of labourers, domestic servants, and 'common' soldiers and sailors, and imputes a subsistence income (44 per cent of the annual wage for labourers and domestic servants) for 'cottagers and paupers' (constituting 23.6 per cent of the population), and for 'vagrants'.

In a similar way, my Domesday population has been divided into two groups (but of markedly unequal size), with a market income for the various land-holding groups (which include the clerics, administrators, military personnel, and court officials who in King's estimates are listed separately), and a subsistence income imputed to the unfree peasants (some of whom in 1688 were tenant farmers and wage-earning workers). Hence, despite the emergence of factor markets and the change in the structure and organisation of the economy over these six centuries, the market/non-market categories in both 1086 and 1688 have been treated in a broadly consistent way.

Prices and real values

It is well known that the most vulnerable part of growth arithmetic is the conversion of estimates in current prices to those in constant prices. The list of problems associated with price deflation techniques is long and well rehearsed – the representativeness of the available prices; the use of shadow prices for goods not marketed; index number problems; fixed weights; quality changes – to name the most important. Yet these problems must be grappled with and adequately evaluated if we are to say anything at all about economic growth. These problems, of course, are not confined to the period before 1688.

Fortunately, most of the quantitative work that has been undertaken on economic change in the Middle Ages has focused upon commodity prices. Indeed, the pioneering work in British quantitative economic history by Thorold Rogers, entitled *A History of Agricultural Prices in England*, was concerned with longrun trends in English commodity prices. This monumental work in seven volumes was followed up by W. H. Beveridge from the 1920s (particularly in *Prices and Wages in England*) and from the 1950s by Brown and Hopkins, T. H. Lloyd, and D. L. Farmer.[37] As a result of these detailed studies, considerable data exist to enable the construction of annual price indexes over at least part of the period since the middle of the twelfth century, for wheat, wool, livestock, and household consumables such as bread, meat, fish, butter, cheese, tea, sugar, malt, hops, textiles, fuel and light. The most complete of these time series is the price of wheat, which is available from 1165, while the prices of wool and household consumables are not available until 1209 and 1264 respectively.[38] Fortunately, the scattered evidence prior to the middle of the twelfth century suggests that prices during the previous two centuries were remarkably stable. In fact, the rapid increase in prices from the 1190s was an unsettling new departure. As Miller and Hatcher claim:[39]

> What evidence there is seems to indicate that this [inflation from around 1200] was something new. Prices of livestock, at least, were apparently remarkably stable during the tenth, eleventh and much of the twelfth centuries. In the reign of Aethelstan oxen were valued at 30*d*; over 150 years later they were reckoned in Domesday Book to

be worth 24*d* and 30*d*; and in the 1160s and 1170s the King seldom paid more than 3*s*. By 1200 he would pay double that amount.

Because of the nature of compound interest over such long periods of time, the likely errors introduced by this assumption will have little effect upon the estimated growth rates.

Although prices increased by a factor of about 19 – or at a rate of 0.49 per cent per annum – over the six centuries after the Conquest, much of this was confined to two major periods of inflation (which in turn experienced considerable annual fluctuation): the 1190s to 1310s, when, for example, wheat prices increased from 2 shillings to over 8 shillings (a fourfold increase in 120 years); and the 1510s to 1650s, when wheat prices increased from 6 shillings to 46 shillings (almost an eightfold increase in 140 years), double the rate of the first period of inflation. As we shall see, these periods of price inflation correspond closely with the two great waves of economic growth during the period 1086 to 1688.

A weighted commodities index has been constructed from the prices of household consumables[40] – which were taken to be representative of that part (82 per cent) of the 1688 national income that was consumed; and the prices of wheat – which was treated as representative of the remaining part (18 per cent) of national income consisting of exports, investment, and government expenditure. The wheat data in Rogers, rather than Beveridge, were employed both because of the longer time period covered by the former and because Rogers' index shows a slightly greater increase in prices and hence a slower rate of growth of per capita income (a factor of 5.75 rather than 5.85). The separate and weighted indexes, which are presented in Table 7.1, imply that household consumables increased by a multiple of 18.9, wheat by a multiple of 21.7, and the weighted index by a multiple of 19.2.

Table 7.1 A weighted price index, England, 1066–1693 (1688 = 100)

Years	*Household consumables*	*Wheat*	*Weighted index*
1066/86	5.3	4.6	5.2
1165/74	5.3	4.6	5.2
1683/93	100.0	100.0	100.0

Sources: See text.

Estimates of living standards

The GDP per capita results are presented in Table 7.2. As one would expect, the differences in the economic condition of England between 1086 and 1688 are striking. The productive capability of the economy had been transformed, even before the impact of the Industrial Revolution. Real national income increased by an unexpectedly high factor of 19.3; population by a relatively small factor of 3.4; prices by a factor of 19.2; real per capita income – the measure of economic growth – by a surprisingly high factor of 5.8; and real per household income – a rough measure of labour productivity – by a factor of 4.1. Clearly, the transformation of the English economy during those six centuries was far higher than even the most optimistic historians have been prepared to suggest. And the implication of the magnitude of these results is that the economy made considerable permanent gains in per capita income between 1000 and 1500, as well as between 1500 and 1700. These results will require a radical reinterpretation of economic change in England and Europe during the last millennium.

Table 7.2 The economic condition of England, 1086 and 1688

Year	*National income*		*Population*	*Rural h/hold heads*	*Price*	*Real per capita income*	*Real per h/hold income*
	Nominal prices (£m)	*1688 prices (£m)*	*(000s)*	*(000s)*	*(1688 = 100)*	*(£)*	*(£)*
1086	0.137	2.63	1 531	275	5.2	1.72	9.55
1688	50.800	50.80	5 136	1 300	100.0	9.89	39.08

Note: Wales has been excluded from the 1688 total on a per capita basis. The 1801 Census was used to calculate the ratio of population between England and Wales, and this was applied to King's estimate of 5.5 million.

Sources: 1086: See text.
1688: Lindert and Williamson, 'Revising England's social tables', *Explorations in Economic History*, p. 389.

THE NATURE OF GROWTH DURING THE LAST MILLENNIUM

Growth during the Middle Ages and the modern era

Economic growth during the six centuries prior to 1700 occurred at a relatively high rate – in the vicinity of 0.3 per cent per annum (see

Table 7.3 Growth rates for England, 1086–1688

	National income (1688 prices)	*Per capita income (1688 prices)*	*Rural household income (1688 prices)*	*Population*	*Prices*
1. Per annum	0.49	0.29	0.23	0.20	0.49
2. Per generation	13.13	7.56	6.15	5.17	13.11
3. Per century	63.80	33.85	26.47	22.35	63.68

Notes: A generation is taken to be 25 years.
Source: See Appendix and text.

Table 7.3). This unexpectedly high figure implies that, on average, real per capita income was compounded by 34 per cent each century or, to put it another way, by 8 per cent every generation (of twenty-five years). Quite clearly, this is a change that would have been perceived by individuals, some of whom lived out their 'three score years and ten', and by families, which possessed even longer memories. Further, it will be shown that there were periods during these six centuries when growth rates were three to six times higher than the average. Therefore, despite the claims of many historians – both quantitative and non-quantitative – medieval populations were keenly aware of economic change, favourable as well as unfavourable.

But how does growth during the Middle Ages compare with modern growth that began with the Industrial Revolution? Table 7.4 indicates that the average rate of growth of GDP per capita for the 600 years before 1688 was almost the same as that for the first half of the

Table 7.4 Growth rates for England, 1086–1987 (real per capita income)

Years	*% p.a.*	*Years*	*% p.a.*
1086–1688	**0.29**	1830–1870	1.4
1688–1760	0.31	1870–1913	1.0
1760–1780	0.01	1913–1950	0.8
1780–1801	0.35	1950–1973	2.5
1801–1831	0.52	1973–1987	1.5

Sources: 1086–1688: see text and Table A.2.
1688–1831: Crafts, *British Economic Growth during the Industrial Revolution*, p. 45.
1830–1987: Maddison, *Phases of Capitalist Development*, p. 6.

eighteenth century, and as much as 83 per cent of that in the second half of the century. In other words, *growth rates during the first half of the Industrial Revolution were only marginally higher than the average achieved for the previous six hundred years.* Even during the latter stages of the Industrial Revolution (1801–37), the 600-year average was not disgraced – it was as high as 61 per cent of the former. Of course, this comparison between the Middle Ages and the Industrial Revolution is based upon an average rate of growth over six centuries. As we know there were prolonged periods of slump (such as the long decline between 1300 and 1450, when growth rates were negative), rates of growth during the long upswings (such as 1000–1300 and 1500–1620) must have been considerably higher than 0.3 per cent per annum. Estimates presented in Table 7.5 suggest that growth rates between 1086 and 1170 were in the vicinity of 0.6 per cent per annum and those between 1490 and 1600 were approximately 1.3 per cent per annum, and even rose to 1.6 per cent per annum between 1490 and 1560. As can be seen from Table 7.4 and Figure 3.2 (see page 86), this compares favourably with British growth rates achieved in the nineteenth and twentieth centuries. While the growth rate of the sixteenth century was enhanced by the reaction of the English economy to a sudden release from the grip of pestilence, so too the growth rate of the mid-twentieth century was enhanced by a sudden release from the grip of world war. We can conclude, therefore, that the growth record of the Industrial Revolution, and even that of the last two centuries, was not at all remarkable in terms of past experience.

Many authors, on the basis of comparative analysis of modern economies, have suggested that the uniqueness of the British Industrial Revolution lies not in the rate of growth, but rather in three distinguishing characteristics: structural change, technological change, and the initiation of a system of self-sustaining growth. This longrun

Table 7.5 Approximate growth rates, England, 1086–1688

Years	*% p.a.*	*Years*	*% p.a.*
1086–1170	0.6	1372–1491	0.1
1171–1280	0.2	1492–1561	1.6
1281–1301	0.02	1562–1601	0.8
1302–1371	-0.7	1602–1688	0.1

Source: See text.

analysis of the growth of England suggests that there is some truth in this claim, but it also suggests that certain qualifications should be noted. First, economic development during the six centuries preceding the Industrial Revolution was not without structural change. Table 5.2 (on page 178) suggests that in 1086 urban population as a proportion of total population was about 8 per cent, whereas by the early eighteenth century this sector had increased by a factor of 2.5 to include as much as 20 per cent of the population.[41] Considerable structural change, therefore, had occurred prior to the Industrial Revolution. The uniqueness of the Industrial Revolution was the speed at which this structural change took place. The second claim about technological change must also be qualified. Major new techniques were introduced into agriculture (water mills, windmills, irrigation works, field rotations, horses rather than oxen and so on), mining, and transport and communications throughout the Middle Ages.[42] But it may well be true that technological change was not as important as organisational change in the Middle Ages. The third claim is clearly wrong as sustained economic growth, even for as long as three centuries, was a major characteristic of pre-modern societies. Economic growth is definitely not a modern invention.

How was it possible to achieve such high growth rates?

Most scholars have found it difficult to envisage rates of growth of per capita income as high as the estimates in this chapter. To my mind, this is because they have failed to draw a distinction between the consumption of perishables on the one hand and of the services of durables on the other, and because they have not recognised that growth in the very longrun proceeds via great waves of economic change.

First, there obviously is a limit to reductions in the level of food, clothing, and shelter, and that is the level of physiological subsistence. In 1086 the bulk of the population was close to that level,[43] as they had always been, and from time to time many were plunged below it owing to widespread harvest failures resulting from natural disasters (floods, droughts and disease) or man-made causes (invasion, war, reprisals) which laid the land waste – a term often invoked in the Domesday record for the northern and western counties. But it was a level, as was shown in Chapter 5, that was no worse than that experienced in India during the mid-nineteenth century. By 1688, the margin between bare

Table 7.6 Components of per capita income, 1086 and 1688

Year	$\frac{Cp}{Y}$	$\frac{Y}{P}$	$\frac{Cp}{P}$	$\frac{Y-Cp}{P}$
1086	0.86	£1.72	£1.48	£0.24
1688	0.50	£9.89	£4.45	£4.45
Factor increase	–	5.8	3.0	18.5

Notes: Cp = consumption of perishables; Y = national income; P = population.
Sources: The Cp/Y ratios from the text and King, *Two Tracts*; and Y/P estimates from the text.

subsistence and the average consumption level was greater,[44] although not by a multiple of six.

The major improvement over these six centuries was in terms of the consumption of services arising from a vastly improved physical environment in the form of better housing (from, on average, rough shelters often shared with livestock – see Figure 5.2 on page 188 – to more substantial, durable, comfortable, and dignified structures); better village and urban facilities (social, health, and educational, as well as economic); better transport and communication facilities, which reduced isolation; and better facilities for storing and transporting food thereby reducing the impact of disasters. The available evidence (see Table 7.6) suggests that the per capita consumption of perishables increased by a factor of three – which may have been largely due to the expansion of the middle class rather than a significant improvement in the consumption standards of the mass of the population – and that the remaining part of per capita income increased by a factor of nineteen. It is the difference in rates of change of the various components of national income that Gould,[45] for example, overlooked.[46] Also it is probably for this reason that Landes' 'guesstimate' of growth,[47] is only a half of my calculation of the growth of per capita income, but equal to my estimate of the per capita growth of the consumption of perishables. None of this implies, however, that there was no growth before 1086; earlier long-term growth resulted not in an increase in living standards of the masses, but in an increase in the surplus of the relatively small but powerful elite. The remarkable redistribution of GDP per capita from a small elite to the great mass of the population in developed societies is the unique achievement of the modern phase of economic growth initiated by the Industrial Revolution.

Secondly, the longrun average of 0.3 per cent per annum for the 600 years prior to 1700 masks shorter-term fluctuations that are more violent than anything experienced in England in the modern period (although not more violent than in Germany or the USSR in this century) because of the force of the exogenous shocks. As we have seen, growth rates during the period 1000–1200 compare favourably with those during the Industrial Revolution, and the rates achieved in the sixteenth century (see Table 7.5 on page 248) are as impressive as those in the nineteenth and twentieth centuries. It is highly likely that the rates in the first half of the sixteenth century are higher than those before the Black Death, in part as a reaction to the negative rates (−1.4 and −0.2 per cent per annum between 1300 and 1370, and 1370 and 1490 respectively) experienced as a result of the shock of the Black Death and the century and a half of pestilence and war that followed. Partly for the same reason, the 1950s and 1960s appear just as atypical in the modern era as the sixteenth century seems in the pre-modern age.

Growth rates of this magnitude for the later Middle Ages and early modern period should not have come as such a surprise, as there is ample evidence that the surplus over and above consumption requirements was growing significantly during the six centuries after the Conquest. First, there is, as already discussed, extensive evidence concerning the growing capital intensity of infrastructure (or capital deepening) in villages, towns, and ports, together with that of the expanding transport facilities, as well as an increase in capital widening with a population that increased from 1.5 to 5.5 million. Secondly, many scholars have emphasised the steady advance of technology during these six centuries.[48] Thirdly, while the expansion of Europe into the rest of the world throughout this period was facilitated by the technical advances occurring in transport and navigation, it relied upon the growth of surpluses over and above consumption requirements. J. R. S. Phillips, for example, has shown that between 1000 and 1500, European influence extended to Greenland, Iceland, North America, the eastern Mediterranean, Cyprus, the Latin empire of Constantinople, and West Africa; and after 1500 there was a large-scale conquest of the Americas involving the migration of Europeans on an unprecedented scale.[49] Fourthly, these growth rates are consistent with the recent findings[50] that the English economy of the eleventh century, being open to market forces, was economically rational, dynamic, flexible, and enterprising – in short, capable of generating sustained economic growth. Finally, and unfortunately, the potentially important evidence on population heights in the very longrun is, as yet, too incomplete to enable us to draw any conclusions. Of course, it could not be employed as evidence for changes in per capita income, but it could theoretically tell us something about changing levels of the per capita consumption of perishables. But to do this it must be representative of the various socioeconomic groups in England. It is to be expected, for example, that the very marked difference in nutritional status between the manorial lords, the free peasants, and the unfree peasants and slaves would lead to a significant difference in average heights between these groups. We would also need to know how long they had been members of the social class in which they died – were they English earls reduced to serfdom or French peasants raised to the aristocracy through valour in battle? The present data on heights for the late Anglo-Saxon period, obtained from the excavation of burial sites, are available in very small numbers (only 78 observations between 850 and 1086, and 174 observations over the thirteenth and fourteenth centuries), are thinly scattered over these periods of time, and are not identified by socioeconomic status. Indeed, it is completely misleading to assemble these data in tabular form, even when the deficiencies are carefully recited.[51]

The sources of growth during the Middle Ages

Although further research is required, particularly concerning changes in capital stock, it is possible at this stage to reach tentative conclusions about the well springs of pre-modern growth. Table 7.3 (on page 247) shows that the rate of growth of population (0.20 per cent per annum) was only 41 per cent of the rate of growth of real national income (0.49 per cent per annum) and 69 per cent of the rate of growth of real per capita income (0.29 per cent per annum). In other words, while the contribution of population to the increase in real national income, and hence to economic development, was important, it was only two-thirds as important as the contribution of non-population forces (as reflected in the growth of per capita income) such as the growth of capital stock (both physical and human), technological change, economies of scale, and reduction in transactions costs associated with the transition of the English economy from feudalism to mercantile capitalism. The longrun relationship between the growth of both population and GDP per capita is shown graphically for the last millennium in Figure 7.2. What is interesting here is the contrast in this relationship between the periods of modern and pre-modern growth. From 1086 to 1688, the rate of growth of GDP per capita was slightly higher than that of population, but after 1688 this relationship was reversed, at least until after the Second World War, with population growing at a significantly faster rate than GDP per capita. This suggests that during the process of modern growth, the trade-off between the increase in population and the improvement in average living standards shifted towards the former. What this appears to mean is that modern growth generated a more equitable distribution of income, by transferring a higher proportion of the national surplus, from the 'ruling elite' to the 'working class', thereby enabling a population to grow at the expense of the living standards, not of the working class, but of the ruling elite. This could have resulted from a transfer of the national surplus between social groups with significantly different average propensities (from a lower to a higher level) to consume.

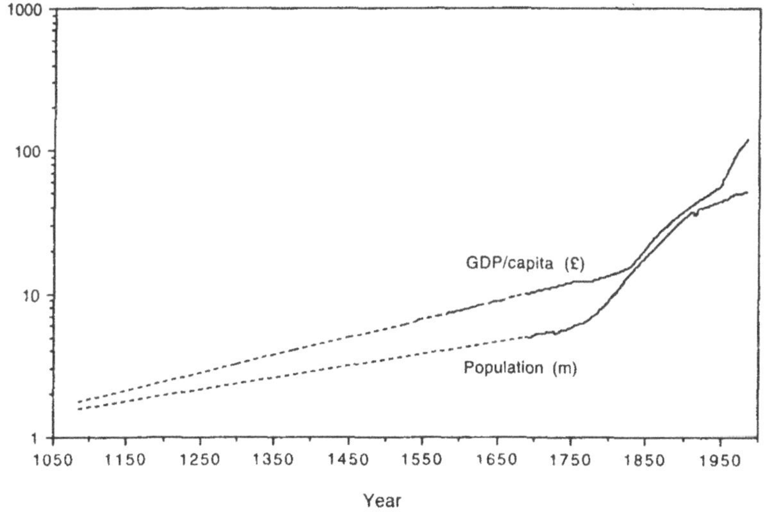

Figure 7.2 The longrun growth of population and GDP per capita, England, 1086–1990

Notes: The curves between 1086 and 1688 describe average rates of change. Log scale on vertical axis

Sources: See text.

This is a fundamentally important conclusion, because it has major implications for the way we view the nature of economic expansion and growth in the pre-industrial period – a period that encompasses most of human experience. It is important for at least two reasons. First, it challenges the overly strong emphasis often placed upon population growth as the dynamic driving force behind economic development during the last millennium.[52] The problem with these intuitive theories is that they are not based upon empirical investigation – no attempt has been made to quantify either the rate of economic change or the contribution of the main sources of development – and tend to be models in search of accommodating events. The empirical results in this chapter suggest that, although population growth may have acted to initiate and terminate 'shorter' surges (even of a century or so in duration) of economic expansion, its direct contribution to longrun economic development during the Middle Ages was not as great as many have claimed. This is not, however, to belittle its achievement. But the great economic achievement – and it *was* great – of the six centuries before the Industrial Revolution was not just a more widespread and intensive colonisation of England (although that did occur), rather it was a more efficient use of its natural and labour resources. This conclusion appears particularly robust because, if anything, I have underestimated the GDP per capita growth rate.

The second, and related, reason for the importance of these results is that it focuses our attention on the forces influencing the growth of productivity. These forces are major players in our drama. The educated guess made by historical economists[53] that any growth that did occur between 1500 and 1700 was largely the result of a reduction in underemployment is not supported by the results in this chapter (see Table 7.3) which show that labour productivity growth (column 3) was as high as 80 per cent of economic growth from 1086 to 1688. We need to ask – although I will not be able to answer it fully here – what accounted for the amazing increase in labour productivity? Modern economists normally think of productivity advances in terms of changes in technology, economies of scale, and the quality of factors of production. Although these forces are important, this approach is limited because it employs the growth accounting approach which is based, as we saw in Chapter 3, upon a comparative static production function model. That approach, by ignoring time, assumes that institutional arrangements are basically static. More recently, some historical economists, namely D. North and R.P. Thomas, have attempted to move the debate beyond the timeless analytical approach of modern economics by considering the longrun growth implications of institutional change.[54] While this has been an important contribution, the emphasis of their work seems to exclude longrun changes in human and physical capital accumulation and those narrowly economic forces driving productivity change; and like all the other studies of this topic it lacks a clear empirical focus, and thereby shares the disembodied nature of the genre. The transformation of Europe from a feudal (or proto-feudal) economy to a mercantile capitalist system greatly improved the efficiency of economic relationships within and between countries as markets evolved further and transactions costs were reduced.[55] At the same time there was a steady increase in human skills and physical capital stock (particularly in transport and communication), and a buildup of appropriate technology, all of which was necessary to support the fundamental changes taking place in England's economic system. Although this must be thought of as a highly integrated process of institutional, physical, and intellectual change, we need to emphasise the importance of the institutional base that economists treat as given.[56] An attempt will be made to give some empirical content to this story in future work.

Non-quantitative economic historians – such as David Landes, John Gould, and Eric Jones – have also emphasised the role of technological change. Jones, for example, has focused upon long-term 'technological drift', which he views as being a function not of economic forces, but of a psychological will to innovate. This obsession to innovate, which because of its alleged pervasiveness apparently requires no economic explanation, is like a spring bubbling constantly to the surface that only fails to nourish economic growth because it is diverted into non-economic uses by an oppressive class of 'arbitrary' warlords. Only after the 'arbitrary' intervention of powerful ruling elites is overcome, by political rather than economic forces, does the upwelling of the will to innovate become translated into an increase in per capita income, which will be reversed if there is a change in the political climate. To my mind, there are many problems with the type of explanation that makes an appeal to non-economic and largely unanalysed forces, not least of which is the fact that there is abundant evidence to show that feudal, and pre-feudal, warlords did *not* act in arbitrary, and hence economically irrational, ways.[57] Also, the structure of inductive models of this type is highly sensitive to the *assumed* rate of growth which, as we have seen, is certainly not valid for England, and is probably not valid for Western Europe either, over the past millennium.

GREAT WAVES OF ECONOMIC CHANGE, AD 1000–2000

It seems quite clear from the above examination of changes in population and average living standards, that the growth process of the English economy passed through a number of long upswings or, as they are called here, great waves of economic change. A computer simulated representation of these great waves for the period 1086 to 1688, which is based upon changes in population, GDP, and GDP per capita, is presented in Figure 7.3. Although this figure has been reconstructed carefully from detailed data concerning the trend, amplitude, and shape of the fluctuations in these macroeconomic variables, it makes no claims for precision in the shorter term.[58] Roughly speaking, the great waves of growth and expansion over the past millennium appear to be three in number: 1000–1300, 1490–1620, and the 1700s to the 1990s and beyond. Each of these upswings was both rapid and prolonged. The first and the third upswings took about 300 years to work themselves out, while the second, which was largely conditioned by an earlier century or so of demographic and military disruption, came to an end after 130 years. As Figure 7.3 shows, these episodes of growth coincide closely with sustained increases in wheat prices, which not only provide greater confidence in the simulated real GDP and GDP per capita results (as they were not part of the simulation procedure), but also suggest that the driving forces behind these great waves are to be found on the demand side. In the absence of the massive exogenous shocks of the fourteenth and fifteenth centuries, it is possible that the middle wave would also have been of approximately 300 years duration. But let me quickly add that while the evidence supports great waves of a systematic nature, it would be purely coincidental if they were also regular in duration.

The first wave of growth and its aftermath, 1000–1490

The simulated economic fluctuations in Figure 7.3, which are determined by a number of solid empirical benchmarks, suggest that there was a strong growth episode between 1000 and 1170 during which population grew at a compound rate of 0.7 per cent per annum and GDP per capita grew at 0.6 per cent per annum – double the rate for the entire six centuries. And there were probably shorter periods within these two centuries of upswing when the growth rate was even more rapid. This expansion and growth in real terms also corresponded to

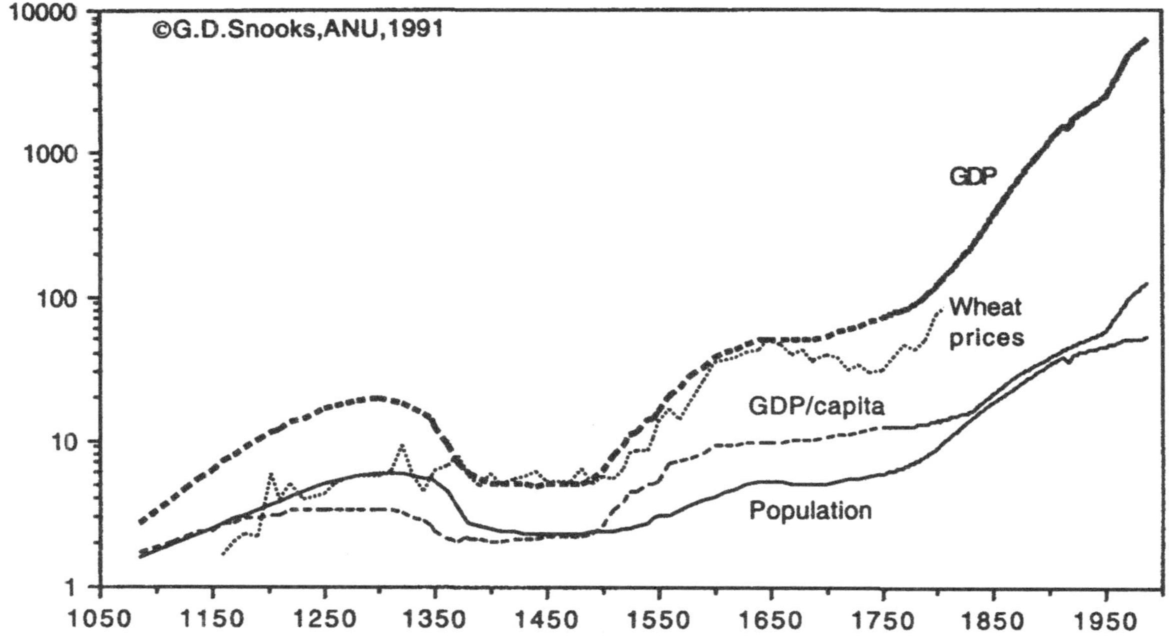

Figure 7.3 Great waves of economic change in England during the past millennium

Notes: Log scale on vertical axis.
Source: 1086–1688, computer simulation, see text; 1688–1990, direct estimates, see Table 7.3; wheat prices from Farmer, 'Some price fluctuations in Angevin England', *Economic History Review*; and Rogers, *A History of Agriculture and Prices in England.*

an increase in agricultural prices, which suggests that demand-side forces were driving the process of economic change in this period. Quite clearly, this great wave of economic change was impressive even in comparison with the one we are currently experiencing.

It is widely accepted that expansion during the twelfth century ended around 1300 owing to the pressure of population on resources at a time when agrarian technology was largely static – a fact that is reflected in the stable wheat yields on Winchester estate manors.[59] Figure 7.3 confirms this interpretation. In fact, these results suggest that growth may have ceased from the middle of the thirteenth century, at a time when population and real GDP continued to expand. The usual explanation is similar to that of the classical growth model, particularly the Malthusian version presented in Chapter 3, in which the growth of population leads to the settlement of progressively less productive land, the emergence of diminishing returns on investment, and a steady decline in the productivity of labour. The effect of this, in the absence of any major new improvements in agricultural technology, is to reduce profits and wages, which in turn leads to a reduction in investment and population expansion until stagnation and even negative growth occurs. At first sight, therefore, the classical model appears far more relevant to the early fourteenth century than it ever was to the early nineteenth.[60] In any case, it would appear that a self-correcting downturn occurred before the impact of the Black Death in 1348 which, as a first approximation, can be treated as largely exogenous. After this dramatic decline, longrun stagnation set in owing to regular and frequent occurrences of pestilence and war (including the Hundred Years War in the thirteenth and fourteenth centuries and the wars with France, Spain, and the Hanseatic League in the fifteenth century) which disrupted local economic activity and international trade. These events combined to cause this great wave of economic change to collapse.

Comparative static interpretations

My interpretation of this wave of economic change, both its powerful upsurge and its 'collapse', differs from the traditional wisdom. Scholars such as Ronald Lee, John Hatcher, M. Postan, and Wilhelm Abel see this episode, as indeed they see the wider period 1250 to 1700, as being dominated by an inverse relationship between *long* swings in population and living standards. The reason for this rather curious view is that they take seriously the Brown–Hopkins real wage rate index. Accordingly, they are forced to develop ingenious, but totally unrealistic, arguments to support untenable empirical relationships. Lee takes a particularly extreme position in this respect. He argues that economic growth is driven by exogenous forces that affect mortality. In this simple model, plague increases mortality, which in turn reduces population and leads to a backward shift of the supply of labour schedule (Figure 7.4 (a)); as the demand for labour schedule is assumed to be fixed (on the heroic assumption that there are no longrun changes in technology, economies of scale, factor proportions and so on), the real wage rate ($\frac{w}{p}$), which is treated as a proxy for per capita income, will rise. In Lee's words:

> We conclude that the real wage and population size of preindustrial England was regulated by an unchanging homeostatic system, in which exogenously varying mortality determined the equilibrium levels. We are left with the impression of great variation in the preindustrial period, but very little structural change prior to the eighteenth century.[61]

Hence longrun growth in this model is generated when population falls, and *negative* longrun growth occurs when population increases. I

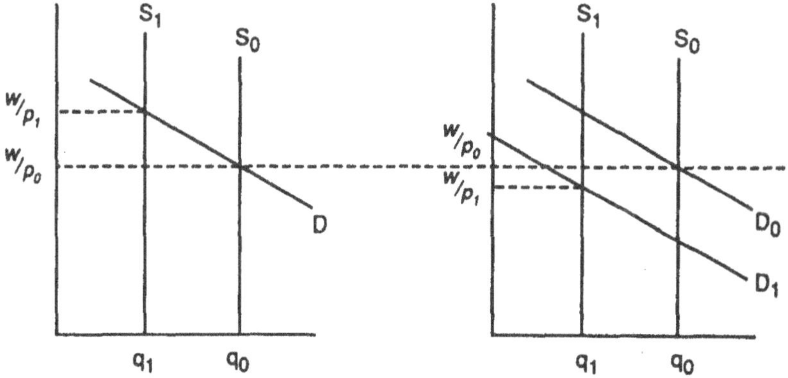

(a) The traditional interpretation (b) A new interpretation

Figure 7.4 Demand and supply determinants of labour earnings following the Black Death, 1348

Source: See text.

cannot think of any conceptual rationale or real world support for such a model, apart from the fact that the author believes the Brown–Hopkins index of real wage rates. If the longrun relationship between population and per capita income is positive after 1700 (as it is shown to be in Figure 7.3 on page 257) why should it be negative before 1700? A further point to note is that, contrary to Lee's claim above, there was, in fact, considerable structural change between 1086 and 1688, with urban population increasing from 8 to 20 per cent of the total. What was driving this structural change if it was not economic growth?

John Hatcher is clearly not comfortable with a model in which growth is only generated by exogenous forces. Yet he feels constrained by the Brown-Hopkins index, and attempts to support the idea of a *longrun* inverse relationship between changes in population and living standards. Referring to the Brown–Hopkins index in conjunction with his population estimates Hatcher concludes: 'It can be seen at a glance that high population coincided with low living standards, and low population with high living standards, and that as population rose so living standards fell, and that as population fell so living standards rose.'[62]

And these are not short-term relationships but rather relationships that persist for some 500 years. In attempting to explain one of these episodes, the Black Death, Hatcher argues:

> There are many good reasons why the reduction in population should have led to a rise in *per capita* output. It is now generally agreed that England, along with many parts of Europe, was suffering from some degree of overpopulation in the early fourteenth century, and that diminishing returns had long since begun to operate in many sectors of the economy. There is widespread evidence of the cultivation of poor soils at this time, of holdings far below the optimum size, and of an abundance of labour which inevitably produced chronic under- and unemployment. The subsequent reduction in population must have led to increased productivity by restoring a more efficient balance between labour, land and capital. The reduction in population must also have led to a sharp increase in *per capita* wealth and consumption. In simple terms, the survivors inherited the property of those who had perished and, when presented with a sudden increase in wealth at a time of recurrent plague and considerable uncertainty, it is not surprising that they chose to spend on a greater scale than their predecessors. Demand was further stimulated by the increasing

> earnings of labourers and peasants; and there is also the possibility that these groups had a greater propensity to consume than landlords and others who suffered a relative reduction in income.[63]

The reasons given by Hatcher fall into two categories. The first part of his argument, about an improvement in productivity resulting from the use of better quality land, is consistent with a move along the demand curve as represented in Figure 7.4 (a). Basically it is a supply-side argument. The second part of Hatcher's explanation is far from clear. He claims that the sudden reduction in population must have increased per capita wealth and consumption, which in turn would have increased aggregate demand, prices and, presumably through its impact on the marginal revenue–product curves for labour (which will move to the right), nominal wages. Unfortunately Hatcher has mixed together arguments about real and nominal wages in this paragraph.

Similar supply-side arguments, that are not always internally consistent, are advanced by scholars such as Abel and Postan. Abel, for example, argues that 'There were not enough men to till the soil; moreover, they demanded higher wages from the people who had been spared by the plague'; and again, 'As the arable shrank the supply of grain dwindled but, on the other hand, although there were also fewer consumers their purchasing power had increased. The many legacies they had inherited, together with the higher wages and prices, had given them a margin bigger than ever before'.[64] Postan wrote about this period in a similar way: 'Increased demand for agricultural labour is difficult to reconcile with what we know of changes on demesnes' and 'that however effective may have been the additional demand for industrial labour, it was not as important a "real" factor behind the movement of wages as the changes in population'.[65] This is the familiar supply-side argument – the type of argument that is required to produce the increase in nominal and real wages shown in the Brown–Hopkins indexes.

The interpretation of Abel, Hatcher, and Postan concerning nominal markets either underplays or overlooks the effect of a massive reduction of population on aggregate demand in the English economy. In a closed economy, during those circumstances, the aggregate demand function will shift inwards at about the same rate as the aggregate supply function. But as the English economy in 1348 was not closed, it is highly likely that aggregate demand began falling at least a year before the aggregate supply function, because of the earlier impact of bubonic plague in Europe. The Black Death, which was first recorded in south-eastern Europe at the end of 1347, had reached Paris by June 1348, London by December 1348, and Edinburgh by the end of 1349.[66] This would have dramatically reduced export demand during 1347–48. After 1348 this export-led reduction in aggregate demand must have been accelerated by a sudden drop in consumption, investment, and government expenditure in England. The effect of this would have been two-fold. First, the resulting fall in product prices – wheat prices did fall slightly in 1348 and more dramatically (37 per cent) in 1349 – would have caused the marginal revenue product curves for labour faced by producers to shift backwards, and for nominal wage rates to fall, at least during 1347–48. But, of course, this tells us nothing about real wages. Secondly, the demand schedule for land would also fall while the supply curve would remain fixed, thereby leading to a reduction in the price of land. The same is true of fixed capital, at least until the capital stock began to deteriorate, and also of moveable goods. Hence there would be no major positive wealth effect as Hatcher and other historians have suggested.

A more likely outcome, based on the available evidence, is that presented in Figure 7.4 (b). As in the first diagram, the labour supply schedule shifts to the left owing to the reduction of population by about 50 per cent in 1348. At the same time the labour demand schedule shifts even further to the left because of the impact of the Black Death on the nature of production and economic organisation, with the result that the real wage falls. The underlying reasons for the leftward shift in the labour demand schedule are four-fold. First, the halving of the economy's size would have resulted in a backward shift along the longrun average cost curve, resulting in diseconomies of scale. These increased costs of production would have been significant, and would have persisted until expansion of the economy was underway again 150 years later. Secondly, the organisation of the English economic system had to be adapted to a scale of operation that had existed 300 years before, and which none could remember. Inevitably organisational costs would have risen. Thirdly, the stock of physical capital would have declined dramatically through non-use and poor maintenance, particularly because of the high rates of depreciation on pre-industrial capital. Also much human capital would have been lost. Finally, there may even have been a degree of technological regress, at least in the first few decades after the onslaught of the Black Death. The same arguments in reverse can be applied to periods of expansion and growth.

Dynamic interpretations

The second theoretical argument against the conventional hypothesis is of a dynamic nature. It is just not plausible to argue that a country experiencing a long and savage downswing in population will experience a sustained increase in real per capita income – sustained, according to Brown and Hopkins, for over a century. Those who subscribe to an argument of this nature appear not to have given any consideration to the dynamics of such an economy. There are a number of fundamental questions that need to be asked. What is the driving force in such a growth process? In particular, what could possibly generate such a sustained increase in living standards? Obviously, it is not the incentive effect of population expansion, because population declined, at first savagely, and then steadily for at least a century. It is hardly likely to be capital accumulation, because a decline in population of this magnitude would have resulted in a steady decline in capital stock. The available evidence reinforces this conclusion.[67] Could it, however, have resulted from a dramatic increase in productive efficiency? Certainly, it is not a result of economies of scale as the size of the economy was halved; and technological change does not appear to have been significant, as Hatcher, among others, refers to 'relatively static technology' during this period.[68] In any case, owing to the massive reduction in population, there was little incentive – and owing to disinvestment there were limited opportunities – for the introduction of embodied technology. Even positive institutional change involving a reduction of transactions costs would have been unlikely in this period – a period when the pressure to adapt to population increase had been dramatically reduced. The upshot of this argument, therefore, is that there is no obvious candidate for a long-term expansion of living standards during this period of population decline. It is more likely that the expansion of real wage rates detected by Brown and Hopkins – with the possible exception of the last two decades of the fourteenth century[69] – is a statistical illusion.

A digression concerning the Brown–Hopkins real wage index

The above scenario is certainly not consistent with an increase in nominal wages in the Brown–Hopkins index of up to two-thirds between 1347 and the early 1350s, or an increase in real wages of 21 per cent between 1347 and 1349, and 42 per cent between 1347 and 1361. But there are further and more important reasons for rejecting the Brown–Hopkins nominal and real wage-rate indexes. Consider what they purport to show. The real wage-rate index shows a marked *increase* during the depressed conditions of the late fourteenth and fifteenth centuries; an even more marked *decrease* during the expansion of the sixteenth century – by which time real wages were below those prior to 1300 – and then a slow recovery through two phases in the early eighteenth and nineteenth centuries until, finally, by the mid-1880s, the former peak in the 1470s had just been exceeded. Think of it: real wage rates are alleged to be higher in 1477 than they were in the late Victorian period over 400 years later. So much for the impact of technological change and the increase in the capital–labour ratio on labour productivity. It implies that the labour supply curve shifted outwards much more rapidly than the demand curve over those 400 years. Also, the dramatic fall in this real wage rate index between 1500 and 1650 contradicts the usual interpretation (by Jones, Maddison, Clarkson, and others) that this period experienced modest growth of per capita income.

The question we need to ask is whether there is something atypical about building workers' wages in southern England. To test this I used the raw price data in Rogers to calculate a more representative index employing agricultural wages and prices, but it suffered from the same problems. It is as if the back of this index, like that of Brown–Hopkins, was broken during the century after 1500. After experimenting with different base years, I was forced to conclude that there are fundamental problems with the Rogers wage-rate data, which will only be resolved by taking a more representative sample of wages from manuscript sources. This, of course, will be a massive undertaking, but it will lead to a major revision of the labour history of medieval England.[70]

The second and third waves of growth

The second and shortest wave of growth during the past millennium began at the end of the fifteenth century and proceeded at a more rapid pace – approximately 1.3 per cent per annum for a century – than did growth during the first episode; so rapid that it grew 2.5 times faster than population (at 0.5 per cent per annum) over the same period (see Figure 7.3 on page 257). A growth rate of this magnitude can compare favourably with anything achieved in England during the nineteenth and twentieth centuries for a comparable period of time. In fact, growth during the first half of the period appears to have been about double that of the second half. And once again this wave of growth was accompanied by an upsurge in agricultural prices, implicating demand-side forces. This episode of economic growth probably owed more to the re-establishment of political order, to organisational changes (and resulting reductions in transactions costs), to the growth of towns, and to the rapid expansion of overseas trade (extending the natural resource base), than to technological change.[71]

Indeed, in terms of its duration and rate of change, this wave of growth appears to have been very much conditioned by the previous one and a half centuries of demographic and military disruption. Without this drawn-out exogenous shock, the growth of GDP per capita to 1620 would probably have extended over a longer period of time, perhaps 200 to 300 years, and would have taken place at a correspondingly slower rate. By the early seventeenth century, however, this amazing wave of growth began to break at a population level approximately equivalent to that in 1300 – about 5.5 million – but at a per capita income level almost three times higher. Following a period of impressive expansion and growth, continued population pressure on natural resources during a period of declining trade (from the late 1610s)[72] appears to have brought economic expansion temporarily to an end. Unlike the downturn after 1300, this period of stagnation was temporary because it did not coincide with an exogenous shock of the proportions of the Black Death. Instead, it was a period in which economic agents were free to respond to the market forces generated by population pressure on natural resources. The response was the application of new techniques of production to agriculture and industry that came to be known as the Industrial Revolution. It would appear that without a major change in agricultural and industrial technology and a transformation of transport and communications, the English population would not exceed 6.0 million. This implies that the economic contribution of the Industrial Revolution should be seen not in terms of the rate of economic change – it was less impressive than that of both the sixteenth century and the tenth and eleventh centuries – but in terms of making it possible to break the nexus between population and natural resources that had harried the Middle Ages.

The third and final phase of expansion and growth, also accompanied by rising prices, began in the mid-eighteenth century – with antecedents stretching back into the seventeenth century – quickly gathering momentum until, by 1800, it was proceeding at a relatively rapid rate. This rate of growth was maintained until the middle of the nineteenth century, after which it began to slow down gradually, and then, after the First World War, quite suddenly. There was an impressive, but brief, acceleration in the 1950s and 1960s, with slower growth thereafter. The interesting point is that during the third phase of expansion, population was able to escape from its low level of 5.5 million in 1700 to reach 50.7 million in 1990 because of the major changes that occurred in technology, infrastructure, and economic organisation from the mid-eighteenth century.[73] It is possible that in the very near future we will be facing a natural resources constraint as serious as that which prevailed throughout the later Middle Ages. If so it may take a change in technology and social organisation as major and as unpredictable as that of the Industrial Revolution to prevent prolonged stagnation and instability.

The process of longrun economic growth

Now that the broad outlines of growth during the last millennium have been charted we need a plausible explanation of these great waves of economic change. Further quantitative work is required before anything definitive can be achieved, but at this stage it is possible to dismiss the main existing analytical model, and to sketch the outlines of a new explanation of this growth process. The existing model was developed to explain the alleged inverse relationship between population change and the real wage-rate index of Brown–Hopkins. This model, particularly in the hands of Ronald Lee, depends critically upon the accuracy of the pattern of their real wage rates. Basically, Lee's model is demographic rather than economic in nature, in the sense that it attempts to explain relationships in the existing data between fertility, mortality, population size, and real wage rates, rather than to provide a behavioural explanation of economic change. Lee's model 'states that fertility is a function of the wage, the wage is a function of population size, and the rate of growth of population is the difference between fertility and mortality'.[74] In this model, population change is driven by exogenous forces that determine mortality, while the increase in per capita income is a negative function of population change. It is a supply-side model based on the assumption that there is no significant shift in the demand for labour schedule owing to the mistaken view that forces which influence labour productivity, such as technological change and increases in the capital/labour and land/labour ratios, were absent throughout the period 1250 to 1700. Thus sustained economic growth occurs when population is decimated by external epidemics and the labour supply schedule shifts to the left, and sustained negative growth occurs when there is an exogenously determined reprieve from pestilence such that population increases and the labour supply schedule moves out to the right. What type of growth model is this? A very strange one indeed, which has no parallel in the modern world. This is not an explanation of an endogenous growth process in which economic man – through his risk-taking, abstinence, and innovation – has any role to play; it is an *ad hoc* system of demographic interrelationships driven by external forces – almost as if human society is the plaything of the gods.

The results presented in this chapter suggest a different model of the growth process during the 600 years before 1700. While it is as yet only a sketch, we have enough detail to see that it will be a very different picture to the traditional one presented above. The fact that per capita income and population increased by factors of 5.8 and 3.4 respectively from 1086 to 1688 – with real GDP and prices increasing by factors of 19.3 and 19.2 – indicate that the labour demand schedule moved outwards far more rapidly than the labour supply schedule. This implies that economic change is an endogenous rather than an exogenous process and that it is driven by demand-side forces. However, it was quite clearly subjected to external shocks which temporarily disrupted this endogenous process.

The picture that is emerging involves the following characteristics. In the first place, economic growth was the outcome of an endogenous process in which the driving force appears to have been economic man, who was involved in applying new ideas to the process of both production and economic organisation. Secondly, there was clearly a *positive* relationship between investment, growth, increases in living standards, population, and prices in the longrun. None the less this does not preclude the possibility of inverse relationships appearing around the turning points of these very longrun fluctuations (see Figure 7.3 on page 257) as a result of Malthusian-type crises. Certainly, there is some evidence for this hypothesis in England around 1300 and again around 1620. But it did not persist for long, as population appears to have adjusted to lower average incomes. Thirdly, the evidence allows for an endogenous process of long growth episodes followed by shorter periods of stagnation and possible negative growth owing to overpopulation and demographic readjustment. It seems clear from the evidence of the mid-seventeenth century that, in the absence of exogenous shocks, the economy can adjust to Malthusian crises. Indeed, such crises may even provide the stimulus for the application of new techniques of production and organisation required to transform society, as occurred during the Industrial Revolution. Finally, this endogenous growth process was subjected to massive external shocks, such as the Conquest of 1066, the Black Death of 1348 and also to prolonged episodes of pestilence and war. The Conquest, together with the subsequent rebellions and reprisals, reduced the scale of the economy through the destruction of population and capital stock, but it only interrupted a 300-year episode of rapid growth for about a generation. The Black Death, however, was more traumatic. This incursion actually reduced the scale of the economy by half and delayed further growth for a century or so but, significantly, it did not prevent its eventual and energetic re-emergence after 1500. Similarly, other external shocks such as wars and invasions have not suppressed the essential momentum of human society that has enabled man to emerge from the natural environment and to create remarkable civilisations.

CHARTING THE PAST AND THE FUTURE

Armed with the above estimates of national income for the eleventh century it is possible to chart quantitatively the dimensions of the economic achievement of the Middle Ages, and to provide a new vision of European growth in the very longrun. Within this quantitative framework the processes of economic change that have operated throughout the last millennium, and will continue to operate long into the future, have become clearer. While there is much work to be done, we can see that natural resources, population, capital formation, innovation, and institutional change are involved in very longrun endogenous processes of economic and social change. These processes have worked themselves out through systematic episodes of economic growth, stagnation and, sometimes 'collapse', of approximately 300 years' duration – episodes that have sometimes been disguised by, and possibly even been conditioned by, the age-old horsemen of the apocalypse called war, pestilence, and death. These are the great waves of economic change that have transformed human society through the millennia. The way has indeed been long and hard.

Economic change is a complex multi-dimensional process consisting of cycles within cycles. Historical economists have recognised, at various times and in various places, the business cycle of 3–5 years; the investment (Juglar) cycle of 7–11 years; the construction–mining

(Kuznets) cycle of 15–25 years; and the long (Kondratieff or Schumpeterian) cycle of 40–60 years. And this book has identified the *very* long episodes of some 300 years' duration. Yet policy economists are only concerned with the first few layers of these cycles, almost as if the deeper underlying forces do not exist. At best, their efforts influence only the ripples on the great waves of change that are carrying us into the future.

It has become clear that economic change in competitive environments is difficult to divert. Even an external event as powerful as the Black Death was not sufficient to destroy the underlying longrun momentum in European society. Certainly bubonic plague reduced the scale of the English economy in the mid-fourteenth century by nearly a half, and for a century or so pestilence and war held economic momentum in check. But as soon as the powerful grip of the pestilence was released and political order re-established, economic growth resumed with great vitality. As this momentum cannot easily be turned aside, let alone brought to a halt as some ecologists would advocate, it seems likely to continue relentlessly into the future. It is a momentum, as we have seen, that is driven by the acquisitive and resourceful nature of man.

8 Time Regained?

To union, and firm faith, and firm accord

(John Milton, 1667)

It is time to return to the major theme of the book. While the technical development of economics over the last few centuries has been impressive, it has been achieved at a considerable cost, the full significance of which is only just emerging. As the gameplayers gained control of the profession's centre stage, the realists were relegated to the wings. In the process, economic theory effectively lost interest in, and contact with, the real world. Economics also lost interest in the big questions that have been, and again will be, vital to the future of man – namely the great waves of economic change that are transforming human society. Theory tells us virtually nothing about these forces. Consequently the dimension gained by the development of economic theory – logical precision – has been outweighed by the dimension lost – time.

An attempt has been made here, in a preliminary way, to address the big questions and to sketch out the very longrun processes of economic change in Europe that have given rise to them – questions and processes that theory has ignored. Much work remains to be done, but at least the initial agenda has been outlined.[1] With the increasing pressure of population on natural resources at the present time there is an urgent need to face these issues. The conviction expressed in this book is that a more realistic approach by the economics profession would help to cast much needed light upon them.

IS THERE A FUTURE FOR A TIMELESS ECONOMICS?

The fundamentally important questions facing us today have emerged from the longrun process of change in human society. For example: are the current high rates of population growth consistent with increasing living standards? Can even the present world population of 5.5 billion be supported by this planet indefinitely at the current standard of living? If not, is it possible to throw population expansion into reverse, as some ecologists demand? Will this require the achievement of zero population growth indefinitely into the future? Can the steady state really ever be achieved, or will the attempt merely lead to instability and collapse? A timeless economics is unable to handle these questions precisely because it has no understanding of the deeper processes of economic change or of the forces that drive them. Indeed, a timeless economics does not even recognise that these processes exist. How can it hope, therefore, to understand what lies ahead? Economics must discover the past before it can probe the future.

A timeless economics, as we have seen, is unable to focus upon processes of change. When it addresses obvious current problems, such as those of the environment, it looks only at outcomes, current and future, and attempts to model the main variables in a comparative static way to reproduce those outcomes. No attempt is made to model processes or to include longrun variables in the analysis. Some applied economists do not even attempt to construct behavioural models, and appear quite satisfied to establish mechanical econometric relationships between variables that provide a fairly reliable basis for short-term prediction – prediction without understanding. At best the focus of a timeless economics is upon the undulating surface of the great waves of economic change. Hence, because economists are blind to the deeper historical forces, when a major change does occur it will take them by surprise.

Had economists existed in the distant past, imagine their surprise and shock when the 300-year surge from the year 1000 broke down around 1300; or when the 150-year surge from 1490 came to an end in the mid-seventeenth century. And will it be any different if, or more likely *when*, the current upswing of 300 or so years, which has been propelling Western society ever since the mid-eighteenth century, comes to an end? No doubt when this great wave of economic growth breaks, timeless economists will be left high and dry. It will be like the explosion of a star deep in the universe, that suddenly bursts onto the night sky – an explosion that took place deep in the past but which, because of the distance between that great event and our time, has only just reached us. Only then, when it is too late, will they give any attention to the pulsars emanating from the past. A warning of what is to come can be provided only by an understanding of very longrun processes of economic change.

As timeless economics is myopic in its focus and deals with outcomes rather than processes, it cannot explain the really essential things. History can. But only with the assistance of economic tools. History can uncover the deeper processes that span past, present, and future. It can show how economies have changed, and are changing, through time. History can demonstrate the importance to the underlying process of change of those variables that economists assume to be constant, such as institutional arrangements. History can explode the myth that the 'growth' theory of economists has anything useful to say about the *process* of change. It can even show, as in the case of the classical economists, that 'growth' theory gives a false impression about the existence and nature of longrun growth. History can, however, demonstrate the positive features of economics. It can explore the universal nature of economic rationality and the role it plays in the deeper forces of change. And it can show that while theory is currently unable to handle growth processes through time, it none the less is very effective in analysing economic systems in the distant past *at a point in time*. This is the value of history.

The future of the economic sciences, including economic history, really matters only because there is no other discipline that has the potential to fully understand the forces of economic change that condition the present and that will determine our future. But because economics has not so far fulfilled this potential, owing to the failure to achieve a *rapprochement* with history, another group of scholars are, quite naturally, demanding to have a determining influence upon public policy. These scholars are from the natural sciences. Unfortunately, although natural scientists are experts in the way the physical environment operates, they have little or no understanding of the way economies operate, or how human society has changed over time. While ecologists quite correctly insist upon the fragility of the natural environment, they do not appear to understand the delicate balance in human society. Indeed, their perspective is not primarily of human society at all, but of the wider ecology. If there is a conflict of interests between ecology and economy, it is likely to be resolved by the natural sciences to the detriment of human institutions. And, like economists, the natural scientists do not understand the deeper forces of change in human society, which they are determined to alter in a fundamental way. If the radical ecologists had their way, critical damage would be sustained by human society. It is essential to recognise that our predicament is man-made and will be resolved, if it is to be resolved at all, more by a knowledge of man in society than a knowledge of the physical universe. That is why the future of economics matters, but only if it can recover the lost dimension of time.

HOW CAN TIME BE REGAINED?

Economics cannot regain time by turning back the clock. The emergence of deductive theory was a necessary step in the development of economics. It is most unfortunate, however, that the process by which this occurred led to the neglect of, and contempt for, history. Central to this process has been the emergence and eventual dominance of the gameplayer in the profession, to the exclusion of the realist. I have attempted to explain this as the outcome of a process of maximising career development by the economic man in the economist. What is required is a *rapprochement* between the gameplayer and the realist; between the theorist and the historian; between the maker of models and the keeper of time. It is absolutely essential in the economic sciences that the time dimension be regarded as the equal of the theoretical dimension. This will require a union, a firm accord, between theory and history.

The partnership of theory and history should encompass research, education, and policy advice. This could be achieved through joint research activities; through undergraduate and postgraduate teaching programmes that contain history as well as theory courses; and through policy-advising teams that include theorists, applied economists, and historians. The nature of the history involved in these joint activities would also be important. It is essential that the time dimension not be limited to the last generation, but that it be extended to include both the longrun and the very longrun. As has been suggested above, the role of longrun analysis is essential if we are to understand, and take into account, the various layers of economic forces of varying longevity that are, and will be, operating in our time and in the time to come.

It is, of course, easy to make recommendations such as these. What we need to know is whether there is any mechanism by which they are likely to be carried out. As argued earlier, the forces that determine such matters within the economics profession have been, are, and will remain centrifugal, owing to the economic man in economists. As economic man generally triumphs over moral man, any appeal to idealism in the economics profession is bound to fail. Possibly the only consideration that will lead economists to heed this call to put their house in order will be an external force that will increase the costs of non-cooperation.

An external force capable of forging greater co-operation between theorist and historian appears to be emerging, owing to the determination of natural scientists to challenge the premier policy-advising role of economists. Although this new force has been gathering strength over the last decade or so it is, as yet, no more than an irritation to economists. It largely takes the form of public name-calling, and has resulted in a modest decline in public esteem. But it is highly probable that the rapidly growing groundswell amongst the general population – particularly the young – over environmental issues may turn irritation into a serious challenge over the next decade. To meet this challenge economics will need to face the big issues of economic and environmental change by coming to grips with historical processes. While this will lead to greater co-operation between theorist and historian, greater co-operation should also be sought with ecologists.

A RESPONSIVE ECONOMIC HISTORY?

While emphasising the importance of history to theory, we also need to recognise the importance of theory to history. The *rapprochement* must involve a fertile two-way interaction. But for this to occur, the theorist and the historian must share a common world view and a common language. The type of history that will satisfy this requirement is what I have earlier called analytical economic history. The hallmark of this approach to history is the use of economic theory to assist in the reconstruction of the past by posing relevant questions, suggesting causal relationships between variables, and postulating outcomes under severely constrained circumstances. But theory is not to be used slavishly; it is little more than a box of tools – albeit powerful – that needs to be employed in a flexible, pragmatic, and disinterested way. If a particular model does not prove useful, then it must be discarded. It should also be stressed that the questions asked must not be limited merely to those suggested by theory. Some of the questions tackled in economic history will very likely be stimulated by scholars in other disciplines, although they will often need reshaping in the light of economic theory. Also while theory has a lot to say about some of the smaller everyday issues – such as the system of production and of markets for commodities and resources – it is silent on the larger longrun issues – such as the nature, course, and direction of very longrun development. The big questions require a more imaginative, and essentially more realistic, approach. But once the larger questions have been posed by history, theory is indispensable in setting out the issue in a workable way. It would be a great mistake, indeed a disaster, if analytical economic history were transformed into economists' history, which is but a pale shadow of the real thing. Economic historians rather than economists must remain the custodians of real time.

The use of quantitative techniques is a second important characteristic of analytical economic history, because these techniques provide access to unused quantitative data sources that are essential in reconstructing the past. An excellent example of this is Domesday Book. Although Domesday Book has been consulted regularly over the last 900 years, the wealth of this data source was inaccessible because quantitative techniques were not brought to bear upon it until the mid-1980s. It had only been employed in a highly selective fashion to support views about feudal England, and about longrun change that are refuted by a systematic analysis of the data.[2] The same is true of major statistical sources in most countries and periods of time. Yet I should emphasise that I am not equating quantitative techniques with econometrics. There are various quantitative techniques that can be applied in history, ranging from the use of data on the prices and quantities of commodities and resources needed to throw light upon the operations of markets; the use of output and input indicators at the industry and firm level to determine microeconomic changes; the estimation of aggregates such as saving, investment, GDP and so on to determine the growth and structural change of economies; the calculation of productivity change to analyse the sources of growth; or the use of econometrics to test simple hypotheses – such as the relationship between tax assessments and measures of capacity to pay – or *more* complex theoretical relationships – such as those described by CES or Generalised Quadratic production functions. Like theory, quantitative techniques are merely tools, with the choice of method being dictated by the nature of the problem. It is interesting that while the smaller and simpler problems are more amenable to the use of econometric techniques, the more difficult, complex, and interesting issues cannot be adequately handled in this way. For example, while the estimation of production functions for the manorial economy of Domesday England can be successfully achieved, more complicated exercises, such as modelling longrun growth processes, have failed.[3]

In advocating the use of quantitative techniques, however, I am not denigrating qualitative methods or sources. Indeed, I am arguing for a broader range of historical skills than are usually possessed by either quantitative or non-quantitative historians. This is also a plea for greater attention to primary sources, both quantitative and qualitative. It is not sufficient for budding (or even ageing) quantitative economic historians to continue to employ *existing* statistical series, such as the discredited Brown–Hopkins real-wage index. On the contrary, it is essential that quantitative economic historians generate new data.[4] The greatest danger facing analytical economic historians is the trap recognised by Cunningham a century ago when he accused Marshall of perverting economic history by taking a narrow and technical approach to the past.

In the first flush of enthusiasm for the American version of economic history – widely known as cliometrics – a number of gameplayers were briefly drawn into the excitement and predictably produced ahistorical analyses of the past. But in the best cliometricians, the realist rather than the gameplayer has always been dominant. Robert Fogel, for example, appears to have swung even closer to the traditional historian's approach over the last decade, to the point where he now views the discovery and use of new quantitative data sources, rather than theory or quantitative techniques, to be the hallmark of cliometrics. While I support the importance of new data sources, I am not willing to relegate theory to a secondary role. Nevertheless, Fogel's point is worth making because there are still some claiming to be historians who approach the past like Cunningham's violating economist.[5] Only the willingness to use time-series data that extend back in time beyond their own generation distinguishes historical economists from applied economists with no interest in time or historical processes. Their view of reality is equally limited.

But just as analytical economic history must avoid becoming a sterile branch of applied economics, so economic and social history – if it is to survive – must avoid becoming an indistinguishable part of social history. To do this it will be necessary to shake off the anti-economic influence of the historicists. While it is important to examine the interaction between economic and social forces in the past, this is unlikely to be successfully achieved unless economic theory is taken more seriously. And that means squarely facing the spectre of economic man. Indeed, unless the descendants of the historicists can differentiate their activities from social historians along methodological lines they will be unable to maintain a separate identity. Both traditions in economic history at least have this in common. Without a clearly distinct methodology both analytical economic history and economic and social history are vulnerable to take-overs from economics and history respectively. If common ground over the use of economics in time could be achieved, the future of the profession would be more secure.

Can time be regained? While theorists and historians are capable of working together, it is not clear that the gameplayers can be persuaded that reality really matters. As long as the rate of return from playing theoretical games is greater than that from exploring reality directly, this union will not occur. But there are signs that the relative rates of return are changing, because it is being increasingly recognised that the games economists play cannot resolve the emerging crisis of a world population seemingly out of control on a planet with finite natural resources. This perception is beginning to affect the valuation placed upon the advice of the gameplayers. It is now entirely possible that in the near future there will be an increasing demand for the services of realists in the determination of policy. And if this happens the economist could be forced into a closer relationship with the economic historian. But the real question is, can time be regained before the profession loses public confidence? Or is economics without time?

Appendix: Alternative Estimates of National Income (1086) and Growth Rates

Because of the different assumptions regarding average household size used in estimating the population of Domesday England in Chapter 5 (see Table 5.2), it is possible to generate alternative estimates of national income and per capita income as well as alternative growth-rate estimates. To test the sensitivity of our results to these different assumptions, alternative estimates of national and per capita income have been provided in Table A.1, and alternative growth-rate estimates have been provided in Table A.2.

Table A.1 National income of England, 1086: alternative estimates

Category	*(1)*	*(2)*	*(3)*
1. National income (£)			
1.1 Manorial income			
1.1.1 Recorded counties			
demesne economy	71 573	71 573	71 573
subsistence economy	47 132	51 306	55 931
1.1.2 Omitted counties	2 932	3 034	2 852
1.2 Burghal income	11 496	10 708	9 151
1.3 Total	133 133	136 621	139 507
2. Per capita income (shillings)	1.916	1.785	1.525

Note: (1) to (3) are based on assumed household sizes of 4.5, 5.0 and 5.5 respectively, and with slaves treated as heads of households.

Table A.1 shows that the estimates for national income fall in the range £133 133 to £139 507 – a difference of 4.9 per cent – while per capita income estimates vary from 1.53 shillings to 1.92 shillings – a difference of 25 per cent. The important question is: what are the implications of these differences for the estimation of economic growth rates? The answer is reflected in Table A.2, which shows that growth rates of per capita income fall in the narrow range 0.28 to 0.32 per annum. The preferred rate was 0.29 per annum.

Table A.2 Growth rates for England, 1086–1688: alternative estimates

	(1)	*(2)*	*(3)*
1. Per annum			
National income	0.498	0.493	0.490
Per capita income	0.280	0.291	0.317
Population	0.217	0.201	0.187
2. Per generation			
National income	13.258	13.130	13.042
Per capita income	7.259	7.561	8.267
Population	5.597	5.173	4.787
3. Per century			
National income	64.540	63.802	63.288
Per capita income	32.352	33.848	37.398
Population	29.541	22.351	20.568

Note: (1), (2) and (3) are growth rates based upon different estimates of population in 1086. A generation has been defined as 25 years.

Notes

INTRODUCTION

1. Goldsmith, 'Roman Empire', p. 288.
2. The term 'collapse' is not meant to imply total destruction, rather a major setback of greater proportions than is implied by the words 'crisis', 'slump', or 'downturn'.
3. Eliot, 'East Coker'. After all, who else can afford to take the attendant risk?
4. D'Amico, *Historicism and Knowledge*, p. 39.

1 TIME LOST

1. This definition is deliberately general because of the range of different views of past and present economists concerning the essence of their discipline. For a brief survey of these views see Groenewegen, '"Political economy" and "Economics"', in *The New Palgrave*, vol. 3, pp. 904–7.
2. On this definition, applied economics (including experimental economics) is a special case of economic history, in that it is concerned with only one point in time, 'the present', whereas economic history is concerned with all points in time including 'the present'.
3. Just as applied economics is a special case of economic history, so too is the comparative approach, as it employs data relating to events that have happened – the domain of economic history.
4. Naturally, the degree to which these problems affected individual economies in the Western world varied.
5. There are some, such as Paul David, who argue that economists and economically literate historians should co-operate to build a new set of models – path-dependent models – that take into account the influence of the past on the present. See David, 'So, how would it matter if "history mattered"?'.
6. Deane, *The State*, p. 36.
7. Sir Richard Stone kindly allowed me to read some of the draft chapters of his forthcoming book on the foundations of an empirical social science.
8. See, for example, Colander and Coats, *The Spread of Economic Ideas*, pp. 27–8 and 33–4.
9. This can be seen in an illuminating discussion of the English Exchequer in the twelfth century; see Poole, *The Exchequer in the Twelfth Century*.
10. Deane, in *The State*, discusses the early emergence of political economy largely in terms of changing supply considerations, including the general intellectual culture (i.e. the development of a scientific method in the

natural sciences) rather than as a response to both demand and supply conditions.

11. The technically-trained economist, therefore, became the modern equivalent of the court soothsayer or magician of earlier times. It is interesting that P. A. Samuelson, when addressing the American Economic Society in 1961, referred to it as 'this occult body' (Samuelson, 'History of ideas', p. 1).
12. The outstanding works in this area are Schumpeter, *Economic Analysis*, Blaug, *Retrospect*, Deane, *Economic Ideas*, and Deane, *The State*.
13. See Deane, *The State*, p. 34; and Appleby, *Ideology*.
14. I have been able to trace the physiocratic equation of 'productive' activity with agriculture and 'unproductive' activities with manufacturing and commerce back as far as Domesday Book in 1086. See Snooks and McDonald, *Domesday Economy*, p. 19.
15. Coleman in *The Economic Past* (p. 17) sees Smith as part of the 'Scottish enlightenment', which also included David Hume, James Steuart, and John Millar who, by using history in their discussions of political economy, became the 'progenitors of what we now call economic history'. But given their rudimentary treatment of history and the fact that there is no clear relationship with the founders of British economic history (Cunningham, Ashley and Clapham), this argument is questionable.
16. See, for example, Becker, *Family*.
17. Coleman, *The Economic Past*, p. 23.
18. Samuelson, 'History of ideas', pp. 8–10.
19. Quoted in Deane, *The State*, p. 88.
20. Foxwell, 'Economic movement in England'; and Ingram, *History of Political Economy*, pp. 35–57.
21. Curiously, Jevons first attracted attention with his empirical and statistical work, particularly on cyclical fluctuations. While he claimed that economic theory should be based upon a deductive method employing mathematics, he thought that economic policy should be based upon applied economics that employed a statistical and historical method, and that the laws of social progress should be based upon a study of economic history. But, to his mind, economic theory was the key to the wider profession of economic studies. See Koot, *Historical Economics*, pp. 26–7.
22. The intellectual climate is discussed in Deane, *The State*, ch. 7. Also see Collini *et al.*, *Intellectual History*.
23. See Koot, *Historical Economics*, pp. 27–31. He tends, however, to take Marshall at face value.
24. Matthews and Supple, 'Marshall on economic history', p. 19. On the originality of Marshall, see Samuelson, 'History of ideas', pp. 10–11.
25. Koot, *Historical Economics*, p. 147.
26. See Matthews and Supple, 'Marshall on economic history', and Koot, *Historical Economics*, pp. 124, 132, 149.
27. Quoted in Deane, *The State*, p. 135.
28. Quoted in Matthews and Supple, 'Marshall on economic history', p. 7.

29. See Morgan, *Econometric Ideas*, p. 5.
30. The main scholars in this tradition in the first half of the nineteenth century included W. G. F. Roscher, B. Hildebrand, and K. G. A. Knies, and in the latter part of the century included A. Wagner and G. Schmoller.
31. See Holmes, *Coleridge*.
32. Mokyr, 'Evolutionary biology'.
33. The changing reputation of these scholars is reflected in the 1894 and 1987 editions of Palgrave. The 1894 entry on Cliffe Leslie, who wore himself out in his militant campaign against deductive economics, ends: 'he will be seen to have taken a leading part in a great work of scientific reform'; whereas in 1987 he was seen as a useful critic of neoclassical economics at a time when 'radical criticism of deductive economics was badly needed'. His reputation has changed from 'reformer' to 'critic'. This reflects the changing view of the economics profession regarding historical–empirical studies.
34. Although Mill emphasised the deductive approach, his wide education inevitably led him to use inductive techniques when examining practical issues, such as the land question in Ireland. He also gave considerable support to the historicist Cliffe Leslie. See Koot, *Historical Economics*, pp. 44–5.
35. Koot (*Historical Economics*, p. 2) goes further when he claims that this school helped the neoclassicists to destroy classical economics. It is clear, however, that the classical school was a century out of date by this time, and was already dead on its feet.
36. Kadish, *Economic History*.
37. Gras, 'The present condition', p. 210.
38. The most recent work on this subject appears inconsistent on this issue. Koot, *Historical Economics*, argues that, apart from Ashley, the English historical school 'did not aim to construct a systematic historicist economics upon their inductive research' (p. 1), but follows this up with the contrary statement that they 'were much more successful in their first goal of laying low the old than they were in their second of building a new edifice' (p. 2).
39. Deane, *The State*, p. 141, quotes a student of Marshall's on the latter point.
40. Offer, 'Retrospect and prospect'; Hayek, 'The London School of Economics'; Koot, *Historical Economics*, ch. 8.
41. A modest attempt in this direction was made by the author in the case of the theory of the growth of the firm; see Snooks, 'The growth process'.
42. Keynes, *General Theory*, pp. vii-viii.
43. Keynes, *Scope and Method*.
44. Keynes, *Probability*.
45. See Morgan, *Econometric Ideas*, ch. 1.
46. Stigler, 'Palgrave', p. 1731.
47. The topics include: economic history, cliometrics, industrialisation, backwardness, Domesday Book (written by F. W. Maitland in 1894!), Chartism, open field system and so on. The scholars include: Cunning-

ham, Clapham, Ashley, Alexander Gerschenkron, Timothy Coghlan (the only Australian economic historian – what of E.O.G. Shann, S.J. Butlin, and N.G. Butlin, who were figures of the first order? No doubt part of the costs of living and working in the antipodes).

48. Palgrave, *Dictionary* vol. II, p. 747.
49. These topics include: economic history, historical school, historical method, induction method, industrial revolution, domestic system, field systems, early English economic history (eleven pages!), factory acts, famine, Domesday Book (the same entry as that included in the 1987 edition), and many more.
50. For example: gilds, villeins, danegeld, knight's service, scutage and so on.
51. See Snooks, 'What should economists be told about the past?'.
52. For example, see Higgins, *Economic Development*, and Meier, *Leading Issues*.
53. See Rostow, *Stages of Economic Growth*.
54. Deane, *The State*, p. 127.
55. Solow, 'Faith, hope, and clarity', p. 40.
56. Snooks, 'It's time'. By old university system I mean the Australian tertiary system before 1988; since then institutes of technology and colleges of advanced education (former teachers' colleges) were encouraged by the Labor government to assume the title 'university'.
57. Hahn, 'Neoclassical growth theory', p. 625.
58. Morgan, *Econometric Ideas*, p. 263.
59. For a detailed discussion see Morgan, 'The stamping out of process analysis in econometrics'.
60. *American Economic Review* (May 1990) p. 438.
61. Smith, 'Experimental methods', p. 242.

2 TIMELY CRITICS

1. I am not concerned here with the various dissenting factions within economics such as the Marxists, neo-Ricardians, post-Keynesians and so on.
2. Actually, Richard Jones (1790–1855) can probably be regarded as the first historical economist in any country, and the first institutionalist critic of Ricardo. See his *Essay*.
3. Koot, *Historical Economics*. My definition of 'historicist' is broader.
4. Leslie, 'The political economy of Adam Smith', in *Essays*, ch. X.
5. Leslie, *Essays*, ch. XIV.
6. Leslie, *Essays*, p. 179.
7. Leslie, *Essays*, pp. 189 and 186–7.
8. La Nauze, *Political Economy*, pp. 98–134, and Groenewegen and McFarlane, *Australian Economic Thought*, pp. 30–5. David Syme was born in Scotland in 1827, studied philosophy in Heidelberg from the late 1840s, followed the Midas trail to California in 1851 and Victoria in 1852, purchased (with his brother Ebenezer) the *Age* in 1856, of which he became editor (on his brother's death) in 1860. He died in 1908.
9. Syme, 'Political economy', p. 212.

10. Syme, *Industrial Science*, p. vi.
11. La Nauze, *Political Economy*, pp. 110–3. This neglect continues to the present – see Koot, *Historical Economics*, p. 221, where Syme is relegated to the tail-end of a footnote!
12. Groenewegen and McFarlane, *Australian Economic Thought*, p. 33.
13. Ingram, 'Address of the president', p. 626.
14. Ingram, 'Address of the president', p. 614.
15. Ingram 'Address of the president', pp. 614 and 617.
16. Ashley, *English Economic History*, vol. 1, Part 1, pp. xi, xii-xiii.
17. Drummond, *Political economy*, p. 21.
18. Ashley's Harvard inaugural lecture is reproduced in Harte (ed.), *Economic History*, particularly pp. 4–9, from which the following quotes are taken.
19. Cunningham, 'Perversion', p. 491.
20. Cunningham,'Perversion', pp. 492–3.
21. Cunningham, 'Perversion', pp. 493–5.
22. Marshall, 'Reply', p. 507 (emphasis added).
23. See Snooks, 'What should economists be told about the past?'.
24. In Chapter 7 I suggest that Rogers did not get all his facts right, particularly those relating to wages.
25. Cunningham, 'Perversion', p. 493.
26. Marshall, 'Reply', p. 510 and 509.
27. Marshall, 'Reply', p. 509.
28. Marshall, 'Reply', p. 511. Maloney's (in *Marshall*) interests differ.
29. Unwin, 'Economic history', p. 40. Historicist is broadly defined.
30. Clapham, 'Economic boxes', p. 312.
31. Clapham, 'Economic boxes', p. 305.
32. Clapham, 'Economic boxes', p. 310.
33. Clapham, 'Economic boxes', p. 312.
34. Clapham, 'Economic boxes', p. 312.
35. Blaug, *Great Economists*, p. 48.
36. Blaug, *Great Economists*, p. 48. Additional comments on this issue were forthcoming in the *Economic Journal* from Robertson (1924, 1930), Sraffa (1926, 1930), Pigou (1927, 1928, 1930), Shove (1928, 1930), Robbins (1928), Young (1928), and even Schumpeter (1928).
37. Some impractical theorising is inevitable, and even essential, so as to explore new ideas that may have a practical outcome in the future. It is a matter of balance. This type of theorising should be the bubble on the ocean rather than the ocean itself.
38. Pigou, 'A reply', p. 465.
39. Clapham, 'Rejoinder', p. 561.
40. Clapham, 'Rejoinder', p. 562.
41. Clapham, 'Economic history', p. 67.
42. Clapham, 'Economic history', p. 67 (emphasis added).
43. Clapham, 'Economic history', p. 70.
44. See inaugural lectures in Harte (ed.), *Economic History*. There were exceptions to the non-economic trends, such as T. S. Ashton.
45. Ehrlich's *Population Bomb* was written after a disturbing visit to India, rather than as a result of an intellectual journey.

46. Meadows *et al.*, *Limits to Growth*, pp. 23–4.
47. One of the co-authors of *Limits*, D.H. Meadows, explains her responses to the unexpected, and highly critical, attention that the book received, in Ehrlich and Holdren (eds), *Cassandra Conference*. For an example of early academic response, see Cole *et al.* (eds), *Models of Doom*. The reaction to radical new ideas is always savage, because of the threat they pose to the establishment.
48. Ehrlich's works include *The Population Bomb; Population, Resources, Environment; Ark II*; and a series of populist pieces, including *How to be a Survivor* and *The End of Affluence*. Suzuki has attracted much attention with his documentaries which have led to *It's a Matter of Survival*.
49. See Ehrlich, *Population Bomb*, pp. 149–51; and *Human Ecology*, pp. 259–66.
50. In *Population Bomb*, Ehrlich was unsure if pollution would lead to the greenhouse effect *or a new Ice Age*!
51. Quoted in Gordon and Suzuki, *Survival*, p. 164.
52. Quoted in Gordon and Suzuki, *Survival*, p. 160.
53. Quoted in Gordon and Suzuki, *Survival*, p. 175.
54. Ehrlich and Ehrlich, *Population Explosion*, pp. 161–2.
55. Ehrlich and Ehrlich, *Population Explosion*, pp. 192–3.
56. Butlin, *Economics and the Dreamtime*.
57. Ehrlich and Ehrlich, *Population Explosion*, p. 163.
58. Ehrlich and Ehrlich, *Population Explosion*, p. 168.
59. Ehrlich and Ehrlich, *Population Explosion*, p. 168.
60. Ehrlich and Ehrlich, *Population Explosion*, p. 181.
61. Ehrlich and Ehrlich, *Population Explosion*, p. 61.
62. Gordon and Suzuki, *Survival*, p. 162.
63. Gordon and Suzuki, *Survival*, p. 181.
64. Gordon and Suzuki, *Survival*, p. 181.
65. Gordon and Suzuki, *Survival* p. 237.
66. These issues are taken up in more details in Snooks, *Doomsday Debate*.
67. See Anderson and Jones, 'Natural disasters'.

3 THE TIMELESS APPROACH

1. See, for example: Higgins, *Economic Development;* Blaug, *Retrospect*; and Hahn and Matthews, 'The theory of economic growth'.
2. See Harris, 'Classical growth models', p. 445.
3. The model presented here is largely that of Ricardo which incorporated elements of the theory of both Smith and Malthus.
4. Jones, *Growth Recurring*.
5. Schumpeter, *Development*, p. 58.
6. Schumpeter, *Development*, p. 64.
7. Schumpeter, *Development*, p. 235.
8. Schumpeter, *Development*, p. 236.
9. Schumpeter, *Development*, pp. 240–1.
10. Schumpeter, *Development*, p. 243.
11. Harrod, 'Dynamic theory', and Domar, 'Capital expansion'.

12. This was first pointed out by Trevor Swan in 'Economic growth', p. 343; but was missed by Solow in 'A contribution'.
13. Hahn, 'Neoclassical growth theory', p. 626.
14. The main exception was France, which experienced higher levels of inflation possibly as an outcome of their system of indicative planning.
15. See Swan, 'Economic growth'; and Solow, 'A contribution'.
16. These articles, which represent only a fraction of the research on growth, are included in the Bibliography.
17. Hahn, 'Neoclassical growth theory', pp. 625–6.
18. The study is N. G. Butlin, *Investment*, which is discussed in some detail in Chapter 4.
19. Barro and Romer, 'Economic growth', p. 1.
20. Romer, 'Long-run growth'.
21. Grossman and Helpman, 'Innovation and growth', pp. 86 and 90.
22. This he takes from Maddison, 'Per capita output'.
23. Romer, 'Long-run growth', pp. 1011–2.
24. Barro and Romer, 'Economic growth', p. 1.
25. Grossman and Helpman, 'Innovation and growth', p. 86. See also Lucas, 'Economic development' and Erlich, 'Development'.
26. Grossman and Helpman, 'Innovation and growth', p. 87.
27. Grossman and Helpman, 'Innovation and growth', p. 86.
28. On the general issue of trade and growth see Kenwood and Lougheed, *International Economy*. The literature on individual open economies is vast; for example, on Australia see Butlin, *Investment*.
29. Grossman and Helpman, 'Innovation and growth', p. 86 (my emphasis).
30. Douglas and Cobb, 'A theory of production'; Solow, 'Technological change'; Wicksteed, 'Laws of distribution'.
31. Domar, 'Measurement'.
32. For a discussion of production functions, from the most restrictive (Cobb-Douglas) to the most general (Generalised Quadratic), and how they can be applied in history, see Snooks and McDonald, *Domesday Economy*, chs 6, 9, and 10.
33. See Snooks, 'Productivity indexes', Snooks, 'Growth and productivity change', and Snooks, 'Innovation'. When the coefficient of variation is high the geometric index is particularly sensitive to the choice of base years and input weights.
34. See Kendrick, *Productivity Trends*.
35. Denison, 'Growth accounting', p. 572.
36. See Grossman and Helpman, 'Innovation and growth; Cornwall, 'Total factor productivity', p. 661; Griliches, 'Productivity measurement'.
37. This is discussed in Snooks, 'Estimating Australian household labour', and Snooks, *Economic Portrait*.
38. Haig, *Capital Stock*, pp 7–23.
39. Denison, *US Economic Growth;* Matthews, Feinstein and Odling-Smee, *British Economic Growth;* Ohkawa and Rosovsky, *Japanese Economic Growth*.
40. For an important contribution to this subject see North and Thomas, *Western World*.

41. Schumpeter, *Economic Analysis*, p. 53.
42. Schumpeter, *Economic Analysis*, pp. 53–4.
43. Schumpeter, *Economic Analysis*, p. 67.
44. Schumpeter, *Economic Analysis*, p. 106.
45. Schumpeter, *Economic Analysis*, p. 54
46. Schumpeter, *Economic Analysis*, pp. 70–1.
47. Schumpeter, *Economic Analysis*, pp. 156–60. Actually, he refers to guides on public administration from the fifteenth century, but overlooks the English *Dialogus de Scaccario* of the twelfth century.
48. Fontenrose, *Delphic Oracle*, examines all surviving written advice from the Delphic Oracle and concludes that it was in the main practical advice rather than prophecy.
49. There were many other oracles throughout the ancient world but this was the most famous.
50. This discussion is based upon Fontenrose, *Delphic Oracle*, and Flacelière, *Greek Oracles*.
51. Flacelière, *Greek Oracles*, p. 20.
52. See Schumpeter, *Economic Analysis*, p. 67.
53. Flacelière, *Greek Oracles*, p. 68.
54. Flacelière, *Greek Oracles*, p. 31.
55. Just as the Greek 'alphabet' was coined from its first two letters – alpha and beta – so 'futhark' was coined from the first six letters of the old runic alphabet. See Elliott, *Runes*. I have substituted 'th' for the symbol ᚦ in 'fu*th*ark'. For an historical evaluation see Loyn, *Vikings*.
56. Bede, *Ecclesiastical History*, book V, ch. 10.
57. It is interesting in this context that the OED definition of 'divine' includes both 'soothsayer' and 'ecclesiastic'.
58. The Venerable Bede is an excellent example. He has left us the best Anglo-Saxon history of this period, misleadingly entitled *Ecclesiastical History*. Also see Stenton, *Anglo-Saxon England*.
59. See Poole, *The Exchequer*; Loyn, *Anglo-Saxon England*; Stenton, *Anglo-Saxon England*.
60. On disposing of the vast stock of wooden tallies on 16 October 1834 in the furnace of the Houses of Parliament, the ancient building was burnt down! See Poole, *The Exchequer*, p. 91.
61. See Clanchy, *England*, pp. 77–82; and Poole, *The Exchequer*.
62. Quoted in Clanchy, *England*, p. 80 (my emphasis).
63. INDECS, *State of Play 6*, pp. 13–14.
64. See for example, Postgate, *Taxation*.
65. There will always be exceptions, and I trust I will not offend them. My concern is not with the exceptions but with the rule.
66. Holmes, 'Indicative planning', p. 781; Meade, *Indicative planning*; Cave and Hare, *Alternative Approaches*.
67. See Estrin and Holmes, *French Planning*; and Holmes, 'Indicative planning'.
68. See Leontief, *Input–output Economics*; Leontief *et al.*, *Studies;* and Stone, 'The accounts of society'.
69. It has been employed by large corporations and government departments to plan for future material and resource requirements.

70. For example, the Prime Minister of Australia in 1965, Mr. R.G. Menzies, said of a proposal to set up indicative planning in Australia that it was unacceptable because it was a 'coercive influence upon governments' (House of Representatives, *Debates*, 1965).
71. A good example of this is the recent applied interest in saving, owing to its relevance to growth, external balance, and ageing. See Bosworth, 'Global decline'.
72. See, for example, Waterman, *Economic Fluctuations*.
73. See Butlin, *Investment*, ch. 6; and Boehm, *Depression*, chs 9–11.
74. Merrett, 'Banking practice'.

4 THE CUSTODIANS OF REAL TIME

1. Coleman, *The Economic Past*, p. 17.
2. The factual basis for this account was obtained from Kadish, *Economic History*; Koot, *Historical Economics*, Coleman, *The Economic Past*, and Coats, 'Disciplinary self-examination'.
3. Toynbee, *Industrial Revolution*, pp. 6 and 7.
4. Toynbee, *Industrial Revolution*, p. 3.
5. Toynbee, *Industrial Revolution*, p. 2.
6. Toynbee, *Industrial Revolution*, p. 3.
7. Toynbee, *Industrial Revolution*, p. 111.
8. Cunningham, *Industry*, vol. I, p. 9.
9. Cunningham, *Industry*, vol. I, pp. 19–20.
10. Cunningham, *Industry*, vol. I, pp. 20–21.
11. Cunningham, *Industry*, vol. I, p. 22.
12. Cunningham, *Industry*, vol. I, p. 26.
13. Gras, 'The present condition', pp. 210 and 213.
14. The present book is a distant echo of these early works, in that it deals both with economic ideas (Part I) and economic circumstances (Part II).
15. Ashley, *English Economic History*, pp. x–xiii.
16. Ashley, 'Economic history'.
17. Koot, *Historical Economics*, p. 106.
18. See the facsimile edition edited by N.G. Butlin *et al.*
19. Other early players included L. L. Price at Oxford, and George Unwin at Edinburgh and Manchester. See Koot, *Historical Economics* for a more comprehensive discussion.
20. Marshall, *Principles*, p. 12.
21. Cunningham, *Industry*, p. 21.
22. Clapham, *Economic History*, p. xi.
23. Clapham, 'Economic history', pp. 69–70.
24. Clapham, *A Concise Economic History*, preface.
25. See Harte (ed.), *Economic History*, pp. xxi–xxii.
26. Gras, 'The present condition', p. 213.
27. I will use the distinctive, if not completely appropriate, name 'cliometrics' to refer to the form of analytical and quantitative economic history that emerged in the USA in the late 1950s and early 1960s.

Alternative terms which include 'econometric history' and 'economists' history' are less appropriate.

28. Gras, 'The present condition', p. 224.
29. Gras, 'Economic history in the United States', p. 327.
30. Coats, 'Disciplinary self-examination'.
31. Dorfman, *Veblen.*
32. See Groenwegen and McFarlane, *Australian Economic Thought.*
33. Such as B. Sheffield, *Observations on the Commerce of the American States* (1783).
34. Gras, 'Rise and development', p. 24.
35. Bolles, *Industrial History of the United States.*
36. Gras, 'Rise and development', p. 26.
37. Drummond, *Political Economy.*
38. Gras, 'Rise and development', p. 29.
39. Aitken, 'The present state', p. 87.
40. Interview by author with N.G. Butlin, 27 April 1984.
41. Aitken, 'The present state', p. 89.
42. Aitken, 'The present state', p. 93.
43. McCloskey, *Econometric History.*
44. McCloskey, *Econometric History,* p. 15.
45. At this conference the author presented one of the early papers to arise from the Domesday Economy Project. See Snooks and McDonald, 'Tax assessments of Domesday England'.
46. Coats, 'Disciplinary self-examination', p. 14.
47. Field, 'The future of economic history'.
48. Conrad and Meyer, *Slavery*, p. 3.
49. Conrad and Meyer, *Slavery*, p. vii.
50. Conrad and Meyer, *Slavery,* p. 3.
51. Fogel, *Railroads,* p. vii.
52. Fogel, *Railroads*, p. viii
53. McCloskey, *Econometric History.*
54. Fogel and Engerman, *Time on the Cross,* vol. I, p. 4.
55. Fogel and Engerman, *Time on the Cross,* vol. I, p. 6.
56. Williamson, 'An interview with Robert W. Fogel', pp. 26–7
57. McCloskey, *Economic Maturity*, p. vii.
58. McCloskey, 'The rhetoric of statistics'.
59. See McCloskey, *Econometric History*, pp. 46–8; and McCloskey, *Essays on a Mature Economy,* p. 3.
60. Williamson, *American Development*, p. ix.
61. North and Thomas, *Western World*, p. vii.
62. Of those listed as students of the Butlins, only Schedvin studied under Syd Butlin, the rest under Noel. Also, Sinclair worked for Noel as a research student between his Masters degree at Melbourne and his D.Phil. at Oxford.
63. See Jackson, 'Growth and deceleration'; Snooks and McDonald, *Domesday Economy*; Snooks, 'Economic growth during the last millennium'; and Part II of this book.
64. For a more detailed discussion see Snooks, 'It's time'.
65. Keynes, 'Jevons', p. 518

66. A letter from Jevons to his sister, quoted in La Nauze, *Political Economy*, p. 31.
67. He has recently attracted the attention of historians of economic thought in Groenewegen and McFarlane, *Australian Economic Thought*, pp. 30–4.
68. Indeed, Shann attacked Syme as 'an apostle of restriction' because of his protectionist views.
69. Coghlan, *Census of 1891 – Statistician's Report*, p. 270.
70. Coghlan, *Census of 1891 – Statistician's Report*, p. 272.
71. For the timeless approach see Eisner, 'Extended accounts', and for a longrun approach see Snooks, 'Estimating Australian household labour', and Snooks, *Economic Portrait*.
72. Clark, 'Coghlan, Timothy (1855–1926)', pp 409–70.
73. See S.J. Butlin, 'An early estimate of Australian national income'.
74. See Mulhall, *Dictionary of Statistics* and N.G. Butlin, *Australian Domestic Product*, pp. 36–7.
75. Coghlan, *Wealth and Progress*.
76. Census of Victoria (Melbourne, 1891) p. 220.
77. See successive volumes of *Wealth and Progress*.
78. For a full discussion of Coghlan's methods, see Arndt, 'A pioneer of national income estimates', pp. 616–25. Arndt had access to Coghlan's working sheets.
79. See these authors in the Bibliography.
80. Butlin, *Australian Domestic Product*, p. 36
81. I am indebted to Bryan Haig for discussions over many years about Coghlan. See also Hicks, *This Sin and Scandal*, pp. 139–46.
82. Coghlan, *Labour and Industry*, p. 1231.
83. Coghlan, *Labour and Industry*, p. 1410.
84. Coghlan, *Labour and Industry*, p. 1417.
85. Coghlan, *Labour and Industry*, p. 1417.
86. Snooks, 'Shann'.
87. National Library of Australia (hereafter NLA) MS643/6, 24/2/1909 (all letters are to his older brother Frank).
88. NLA MS643/14, 4/6/1909.
89. NLA MS643/16, 27/7/1909 (written from the Fabian summer school in North Wales).
90. NLA MS643/11, 10/5/1909.
91. NLA MS643/5, 16/2/1909.
92. Shann, *Economic History*, p. 385.
93. Shann, *Economic History*, pp. 446–7.
94. See Stone, '1929 and all that'. At his resignation from the public service in 1984 John Stone was Secretary to the Treasury. Ironically, he later became a Senator representing the National Party that strongly supports rural subsidies.
95. See Snooks, 'Shann', p. 575. This theme also appears in Shann's other writings, especially *Bond or Free?*
96. The failure to realise this has led to a gross undervaluation of Shann's works.
97. This theme of early settlers struggling against an alien environment was

initially developed at length in Shann's *Cattle Chosen*, about the Bussell family in the south-west of Western Australia in the early-to-mid-nineteenth century.

98. See Snooks, 'Orthodox and radical interpretations', p. 2.
99. For outlines of their work and lives see: N. G. Butlin, 'Fraternal farewell; and Snooks, 'In my beginning is my end'.
100. See Arndt, *A Course through Life*, p. 18.
101. S. J. Butlin, *Foundations*, pp. v, vi.
102. S. J. Butlin's review of Coghlan's *Labour and Industry* in *Australian Economic History Review*, vol. 11, no. 1 (1971) p. 69.
103. S. J. Butlin, *Foundations*, p. 1.
104. As S. J. Butlin said in *Foundations*, p. v, the 'two most impressive contributions to the theory of the economic development of capitalism' were those by Marx and Schumpeter.
105. For a more detailed discussion see Snooks, 'In my beginning is my end'.
106. This work began appearing in preliminary form in the mid-1950s soon after his arrival back in Australia: see N. G. Butlin and de Meel, *Public Capital Formation* (1954) and N. G. Butlin, *Private Capital Formation* (1955); and N. G. Butlin, 'Shape of the Australian economy' (1958). It appeared in final form in N. G. Butlin, *Australian Domestic Product* (1962); and Butlin and Sinclair, 'Australian gross domestic product' (1986). His other major works include: *Investment* (1964); *Government and Capitalism* (with A. Barnard and J. J. Pincus) (1983); *Original Aggression* (1983); and various edited volumes from the Botany Bay Project of the mid-1970s.
107. Noel Butlin began his national accounting career by reworking, with H. W. Arndt, Timothy Coghlan's early national income estimates for NSW. See Butlin and Arndt, 'National output, income and expenditure of NSW, 1891'.
108. A variation of the Butlin method was subsequently employed to generate regional estimates of GDP for both Western Australia and South Australia. See Snooks, 'Regional estimates of GDP', pp. 536–53; Snooks, 'The arithmetic of regional growth', pp. 63–74; Snooks, *Depression and Recovery* ; and Sinclair, 'Gross domestic product'.
109. With Heinz Arndt and Kingsley Laffer, Noel Butlin established the Fabian Society of New South Wales. In 1947 he was co-author with H. W. Arndt of a widely circulated (40 000 copies) pamphlet entitled *The Case for Bank Nationalisation* (Fabian Society) which was used as propaganda by Prime Minister Ben Chifley in an unsuccessful attempt to nationalise Australian banks.
110. Butlin, *Investment*, p. xiii.
111. Economic History was not separated from Economics in IAS, ANU, until 1962.
112. See Snooks, 'Orthodox and radical interpretations', p. 4.
113. See N. G. Butlin and Gregory, 'Swan'.
114. Butlin, *Investment*, p. xiii.
115. Butlin, *Investment*, p. xiv.
116. Butlin, *Investment*, pp. xiv–xv.

5 ECONOMIES 'LOST' IN TIME

1. E. L. Jones, *Growth Recurring*, pp. 30–1.
2. Fenoaltea, 'Authority'; Fenoaltea, 'Risk'; McCloskey, 'Open fields'; and McCloskey and Nash, 'Grain storage'.
3. Farber, 'Northern Babylonia'.
4. See Johns, *Assyrian Doomsday Book*; and Postgate, *Taxation*. I am currently investigating the possibility of using the quantitative data on Assyrian clay tablets to extend the scope of the present study.
5. See Finley, *Ancient Economy*; Goldsmith, 'Roman Empire'; and Clark, *Economic Progress*; and Levi, *Rule and Revenue*, ch. 4.
6. Attempts to estimate the national income of the Roman Empire have been made by Clark, *Economic Progress* and Goldsmith, 'Roman Empire'.
7. Darby, *Domesday England*, p. 223.
8. Lennard, *Rural England*, pp. 25–6.
9. Stenton, *England*, p. 634.
10. Darby, *Domesday England*, pp. 57–94.
11. Stephenson, *Medieval Institutions*, pp. 184–205.
12. Hamilton, *Inquisitio*, p. 97.
13. See Stenton, *England*, p. 652; and Galbraith, *Making of Domesday Book*, p. 36.
14. Postan, *Medieval Economy*, pp. 90–4.
15. Loyn, *Anglo-Saxon England*, p. 161; Postan, 'Agrarian society', pp. 573–4.
16. Loyn, *Anglo-Saxon England*, p. 334.
17. Postan, 'Agrarian society', p. 573; Campbell, 'Agricultural progress'; and Campbell, 'Arable production'.
18. Loyn, *Anglo-Saxon England*, p. 160.
19. Darby, *Domesday England*, pp. 302–9 and 364–8.
20. Darby, *Domesday England*, p. 293; and Miller and Hatcher, *Medieval England*, p. 9.
21. Darby, *Domesday England*, pp. 261 and 309–10.
22. See Snooks and McDonald, *Domesday Economy*, chs 5 and 6.
23. Ballard, *Inquest*, pp. 152–5.
24. Snooks and McDonald, *Domesday Economy*, pp. 108–12.
25. Darby, *Domesday England*, pp. 87–91. No attempt has been made, as Darby has done somewhat unconvincingly, to guess the magnitude (5 per cent) of omissions (if any) of the rural workforce in the Domesday text. The absence of recorded labourers on holdings that produced income in Domesday Book does not necessarily mean that there were any omissions, because it is possible that these holdings (which are much smaller than the average – ranging from 15 to 25 per cent) were worked using labour from adjacent manors. Also, the proportion of holdings in this category (in value terms) was small – generally less than 2 per cent. In any case, if population is underestimated, the resulting calculation of per capita income will bias the rate of economic growth downwards.
26. Darby, *Domesday England*, pp. 302–9.

27. Darby, *Domesday England*, pp. 87–8, 92; Titow, *Rural Society*, pp. 57–9, 89; Maitland, *Domesday Book*, p. 437; Ellenby, *Anglo-Saxon Household*. Hatcher, *English Economy*, p. 68, has suggested a higher range of 1.75–2.25 million, owing to a larger household count than a larger household size.
28. Jones, *Growth Recurring*.
29. Jones, *Growth Recurring*, p. 133.
30. See Snooks and McDonald, 'Tax assessments', pp. 367 and 371; and Snooks and McDonald, *Domesday Economy*, pp. 62–4, 66, 68, 73.
31. Dyer, *Middle Ages*, p. 46.
32. Miller and Hatcher, *Medieval England*, pp. 151 and 161.
33. See Snooks and McDonald, *Domesday Economy*, ch. 7; Snooks, 'Rational calculation'; and Chapter 7 above.
34. Dyer, *Middle Ages*, pp. 114–5.
35. Anderson, 'Household structure', p. 235.
36. Postan, *Essays*, pp. 150–85; and Razi, *Life*, pp. 43–5, 130–1.
37. See Pyke, *Food*; Braudel, *Civilisation*; Aymard, 'Nutrition'; Carpenter, 'Wheat proteins'; de Vries, 'Crop yields'; Deatherage, *Food*; Taylor, *Nutrition*; Widdowson, 'Extraction rates'; Butlin, *Original Aggression*. Southeast Asian experts have assured me that the calorie intake in some parts of this region is even lower than the figures quoted in this paragraph.
38. Caprihan, *Hunger*.
39. Schofield, *Village nutrition*.
40. Dyer, *Middle Ages*, pp. 151–6.
41. Butlin, *Original Aggression*, pp. 166–8.
42. Dyer, *Middle Ages*, p. 160.
43. Dyer, *Middle Ages*, p. 159.
44. Cipolla, *European Society*, p. 92.
45. Fogel, 'Nutrition', pp. 86–7.
46. Dyer, *Middle Ages*, p. 153.
47. See Dyer, 'English diet'; Dyer, 'Harvest workers'; Hilton, *Medieval Society*.
48. See Maitland, *Domesday Book*, p. 44; and Farmer, 'Price fluctuations', p. 37.
49. Cipolla, *European Society*, pp. 30–1.
50. Cipolla, *European Society*, p. 32.
51. Calculated from data in Dyer, *Middle Ages*, p. 176, taking into account price changes from 1086 to 1348.
52. Cipolla, *European Society*, p. 31.
53. The expected relationship between the 'wages' of free and unfree peasants, given a reasonable degree of mobility, will depend upon the prevailing degree of peace and security. In a time of peace we would expect the average wages of free peasants to be lower than those of unfree peasants, owing to the non-pecuniary benefits of freedom (or disamenity of bondage). But in a time of insecurity, which was more characteristic of England in this period, the security that a lord could offer would outweigh the disamenity of bondage. As Loyn (*Anglo-Saxon England*, p. 196) has

said of the Anglo-Saxon period: 'In an age of peril it was indeed natural for men to seek lords and lords to seek men'. In these more typical circumstances, the wages of free peasants will probably be higher than those of unfree peasants. This is not in doubt for the period of oppression under William the Conqueror.

54. See Domesday Book, Buckinghamshire, folio 148b, for example, where a dispossessed free peasant was recorded as occupying his former land '*graviter et miserabiliter*' – 'in heaviness and misery'.
55. Campbell, 'Concentrated urban demand', pp. 16–17.
56. Darby, *Domesday England*, p. 89.
57. Snooks and McDonald, *Domesday Economy*, p. 19.
58. Postan, *Medieval Economy*, p. 239.
59. See Darby, *Domesday England*, pp. 303–8. The pioneering historian of British medieval population is Russell (*Medieval Population*, pp. 45–51); for an evaluation of his work see Postan, 'Agrarian society', pp. 561–2; and Miller and Hatcher, *Medieval England*, p. 29.
60. This can be seen by inspecting the Domesday folios, which are arranged by county, and within each county by tenant-in-chief. Manors held by tenants-in-chief were scattered widely, both between and within counties.
61. Snooks and McDonald, *Domesday Economy*, p. 21.
62. These assumptions are: an average household size of 5.0; slaves were recorded in the same way as the rest of the population, that is, by head of household.
63. This comparison is based upon the gold equivalent GDP per capita index estimated by Goldsmith, 'Roman Empire', p. 280: 'In terms of gold, which was the base of the monetary system of the early Roman Empire, its national product per head of about four aurei, equal to 31g . . . was higher than those of India and Brazil in the mid-nineteenth century (14g and 27g), but substantially lower than those of England in 1688 (over 70g) or of France and the United States in 1820 (about 100g).' As England's GDP per capita in 1688 was equivalent to 70g of gold, in 1086 it would have been equivalent to 12g.
64. Especially Hatcher, *English Economy*, p. 68; Miller and Hatcher, *Medieval England*, p. 29; and Postan, 'Agrarian society'.
65. Smith, 'Demographic developments'.
66. See also Snooks and McDonald, *Domesday Economy*, ch. 6.
67. Darby, *Domesday England*, pp. 260–5.
68. Darby, *Domesday England*, pp. 266–9.
69. Miller and Hatcher, *Medieval England*, pp. 79–83; Postan, *Medieval Economy*, pp. 185–90.
70. Loyn, *Anglo-Saxon England*, p. 96; Miller and Hatcher, *Medieval England*, pp. 10, 79–80.
71. Geld in 1083–4 was about 16 per cent of the annual values of the average manor in Essex, and feudal dues appear to have been at least as large as this. See Chapter 6.
72. Ideally, we would like to have detailed data on local trade through fairs, together with that on consumption, storage, and inputs of various urban industries. Such data, unfortunately, are not available.

73. Abel, *Agricultural Fluctuations*, p. 5. For a discussion of the operation of the market in the later Middle Ages, see Dyer, 'The consumer and the market'.
74. United Nations, *Tropical Africa*, pp. 9 and 19; and Chandavarkar, *Staff Papers*, pp. 679 and 685 (both quoted in Goldsmith, 'Roman Empire').
75. Goldsmith, 'Roman Empire', pp. 274–5.
76. The separation of the demesne and dependent peasant sector is conceptual rather than geographical as strips were scattered throughout the manor.
77. Loyn, *Anglo-Saxon England*, p. 196.
78. One of the most overwhelming expressions of Norman power and intimidation is Durham cathedral – beautiful but dominating. Compare it with the equally beautiful, but uplifting, Gothic cathedral at York.

6 HUMAN MOTIVATION THROUGHOUT TIME

1. I counterpose moral man with economic man, because the English word 'moral' comes from the medieval Latin word *moralis* meaning 'relating to custom'.
2. See Marshall, 'Reply'; McCloskey, 'English open fields'; and McCloskey and Nash, 'Corn at interest'; and North and Thomas, *Western World.*
3. I prefer to define the objective function of economic man in terms of material self interest rather than in terms of utility which can be extended to the point of meaninglessness. In this form the hypothesis has more explanatory power and is easier to verify or falsify.
4. See Deane, *The State*, ch. 4.
5. See Arrow, 'Rationality'; and Simon, *Models of Man.*
6. Simon, 'Bounded rationality'.
7. Simon, *Bounded Rationality.*
8. Simon, *Bounded Rationality*, p. 267.
9. Johns, *Assyrian Doomsday book.*
10. See Bridbury. He reviewed *Domesday Economy* in the *Economic Record*, 63 (December 1987). He also supports our view that geld was an *income* tax.
11. See Jones, *Miracle*, p. 85; and Jones, 'Transitions', pp. 56–7, 61.
12. Auerbach, 'Taxation of income', p. 604.
13. See Seligman, *Taxation*; and Postgate, *Taxation.*
14. Painter, *Studies*, pp. 74–8.
15. Postan, *Medieval Economy*, p. 192.
16. Round, *Feudal England*; and Galbraith, *Making of Domesday Book*, p. 42.
17. Green, 'Danegeld', pp. 241–2.
18. Round, *Feudal England*, p. 45.
19. The 1986 edition was revised in the light of Snooks and McDonald, 'Tax assessments'; Snooks and McDonald, 'Manorial income'; and Snooks and McDonald, 'Statistical analysis'.
20. Darby, *South-west England*, p. 17.
21. Jones, 'Transitions', p. 61 (my emphasis).

22. Of his surprisingly long life of sixty years, William the Conqueror spent more than fifty years in continual warfare.
23. The diagnostic tests for functional form are described by Godfrey and Wickens, 'Testing', and for heteroskedasticity by Pagan and Hall, 'Diagnostic'.
24. Snooks and McDonald, 'Tax assessments'.
25. Leaver, 'Five-hides'.
26. Snooks and McDonald, 'Mistaken identity'. The paper remains unpublished because the journal that published our original article together with a number of comments upon it, decided to prematurely terminate the debate.
27. Maitland, *Domesday Book*, p. 544.
28. See, for example, Darby, *Domesday England*, pp. 220–1 and Darby and Maxwell, *Domesday Geography*, for example, pp. 43, 125, 199, 302–3 and 352.
29. Hollister, *Military*, p. 13.
30. The literature is undecided as to whether mill renders are included in the values. These results are important because they suggest that the renders were *not* included, probably because of a physiocratic-type view that wealth could only be created from the land.
31. Snooks and McDonald, *Domesday Economy*, ch. 2.
32. Space limitations prevent a discussion of the various theories of the manor. A selection of works on the manor in theory includes: Aston, 'The manor'; Postan, *Medieval Economy*; Postan, 'Feudalism'; Domar, 'Slavery'; North and Thomas, 'Manorial system'; and Fenoaltea, 'Authority'.
33. Titow, Review of *Domesday Economy*, p. 303.
34. See Denny, 'Functional forms'.
35. See Sato, 'Two-level'.
36. For a more detailed account of the argument in this subsection, see Snooks and McDonald, *Domesday Economy*, pp. 105–7, and for a more technical discussion of the Sato function, see pp. 197–201.
37. A similar range of resources is available for Norfolk and Suffolk (and, in more fragmented form, for the counties in Exon. Domesday), but the complexities, due to the influence of Danish settlement, make it very difficult to apply production functions.
38. Sir Richard Stone, who has done similar work on Gregory King's estimates, was impressed with the quality of these results. See Stone, 'Public finance', p. 31.
39. A detailed discussion can be found in Snooks and McDonald, *Domesday Economy*, pp. 108–12.
40. Darby, *Domesday England.*

7 ACCOUNTING FOR THE VERY LONGRUN

1. A measure of the truth of this statement will be the strength of the resistance of reviewers to the growth rate estimates for the 600 years before 1700 presented in this chapter, despite the robustness of the

results, the sensitivity tests conducted, and the various independent checks made.

2. At least until Snooks and McDonald, *Domesday Economy*.
3. Results from this research were first presented in Snooks, 'Last millennium'.
4. Deane, 'Early national income estimates', pp. 3 and 6, (my emphasis).
5. Crafts, *British Economic Growth*, p. 9. My emphasis in following quote.
6. Pollard and Crossley, *Wealth of Britain*, p. 11.
7. Jones, *Growth Recurring*, pp. 30–1.
8. Snooks and McDonald, *Domesday Economy* pp. 32–6.
9. Cipolla, *European Society*, p. 209.
10. Miller and Hatcher, *Medieval England*, pp. 25–6.
11. Estimates of national income for the Roman empire can be found in Goldsmith, 'Roman Empire' and Clark, *Economic Progress*. These estimates are based upon far less comprehensive and reliable data than that in *Domesday Book*.
12. See Kuznets, *Modern Economic Growth*, p. 139; Rostow, *Stages of Economic Growth*, ch. 2; Maddison, *Capitalist Development*, pp. 6–7; Persson, 'Labour productivity'; and Lee, 'Pre-industrial England'. The recent use of terms such as 'intensive' and 'extensive' growth are regrettable because of the inevitable lapse into the ambiguous use of the diminutive 'growth'. I will use 'economic growth' to mean, as economists have always meant, changes in real per capita income, and 'economic expansion' to mean increases in real GDP.
13. Persson (*Pre-industrial Economic Growth*, p. 139) has attempted to estimate pre-1700 English growth indirectly by employing a simple model that depends basically upon a precise and stable relationship between changes in GDP per capita and urbanisation. While this attempt is interesting and imaginative, it is flawed, and generates growth rate 'guesstimates' for 1000–1300 (0.1–0.24% p.a.) and 1500–1700 (0.28–0.34% p.a.) that are far too low – being only a fraction of my direct estimates (see Table 7.5) for 1086–1200 (0.6% p.a.) and 1490–1600 (1.3% p.a.), and even less than the *average* (0.3% p.a.) for the entire period 1086–1700 using direct estimation procedures. Had Persson been able to measure the negative growth from 1300–1500 – but his indirect method cannot handle this – the modest positive growth in 1100–1300 and 1500–1700 would have been cancelled out and the average rate for the six centuries before 1700 would have approached zero. Hence his work is really part of the conventional wisdom that views economic change in the pre-industrial world in terms of zero-sum fluctuations. Indeed, his concluding chapter is entitled: 'Why have growth rates been so low until recently?'.

    Persson's method is flawed because, in the absence of adequate indirect data required to estimate a complex model that could cope with regular changes in its many parameters, he is forced to adopt a very simple model based upon the assumption of a precise and stable relationship between agricultural productivity (his proxy for GDP per capita) and urbanisation. This model embodies a large number of very strong assumptions:

(a) That urban productivity growth is the same as agricultural productivity growth.
(b) That landowners do not change residential location between country and city over the very longrun.
(c) That the location of industries producing non-agricultural (that is, 'urban') goods does not shift between country and city over time.
(d) That international trade does not result in specialisation according to comparative advantage. But, as we know, England specialised in the production of products for export, and it imported tradeable urban goods. Other countries such as Italy did the reverse. Hence in England GDP per capita growth proceeded much more rapidly than urbanisation prior to 1700; and in Italy the reverse occurred.
(e) That the consumption function is stable over long periods of time.
(f) That the expenditure patterns between imported and local urban goods are stable over time and the same for all socio-economic groups (and hence unresponsive to changes in income distribution).

While Persson is aware of some (even all) of these difficulties, and does attempt to control some of them, the 'data' he used are no more than 'guesstimates'. It is not surprising, therefore, that his imaginative attempt is wide of the mark. But it was worth the effort.

14. Lee, 'Pre-industrial England'; and Maddison, *Capitalist Development*, p. 6.
15. Maddison, *Capitalist Development*, pp. 6–7.
16. Brown and Hopkins, *Perspective*, pp. 19 and 62.
17. Gould, *Economic Growth*; Landes, *Prometheus*; Komlos, *Habsburg Monarchy*.
18. Gould, *Economic Growth*, pp. 35–9.
19. Brown and Hopkins, *Perspective*.
20. Komlos, *Habsburg Monarchy*, p. 221.
21. Landes, *Prometheus*, p. 14.
22. Jones, *Growth Recurring*, p. 4.
23. Jones, *Growth Recurring*, p. 38.
24. Jones, 'Recurrent transitions', p. 45.
25. Jones, *Growth Recurring* and 'Recurrent transitions'.
26. Jones, *European Miracle*.
27. Wrigley and Schofield, *Population*.
28. Hatcher, *English Economy*, pp. 68–9; Smith, 'Demographic developments'; and Chapter 5 above.
29. Hatcher, *English Economy*, p. 71 (Figure 2); and Lee, 'Pre-industrial England'.
30. I am currently attempting to quantify the nature of economic change on an annual basis by constructing a series of proxies for per capita income.
31. Miller and Hatcher, *Medieval England*, ch. 2.
32. Jones, *Agriculture*.
33. See Deane, 'Early national income estimates'; and Lindert and Williamson, 'England's social tables'.

34. Crafts, *British Economic Growth*, pp. 14, 17, 65; and Lindert, 'English occupations'; and ch. 3 of Stone's forthcoming book.
35. See Stone, 'War end', 'Consumers' behaviour' and 'Public finance'.
36. This is a limiting convention, and I am conducting a separate project aimed at measuring household activity in the longrun. See Snooks, 'Household labour services', and *Economic Portrait*.
37. In addition to works already cited see Lloyd, 'Medieval England', and Farmer, 'Angevin England'.
38. See Farmer, 'Angevin England'; Lloyd, 'Medieval England'; and Brown and Hopkins, *Perspective*.
39. Miller and Hatcher, *Medieval England*, p. 67.
40. Brown and Hopkins, *Perspective*.
41. See Wrigley and Schofield, *Population*.
42. Jones, *Miracle*; Cipolla, *European Society*; and Mokyr, *Riches*.
43. Postan, *Medieval Economy*; Miller and Hatcher, *Medieval England*; Bolton, *English Economy*; Dyer, *Middle Ages*.
44. Crafts, *British Economic Growth*; Jones, *Agriculture*.
45. Gould, *Economic Growth*, pp. 34–42.
46. This conclusion appears to have important implications for a measure of living standards that has sky-rocketed in popularity in the last few years – the physical characteristics of the population. Average heights may be a reasonable proxy for the changing average consumption of perishables by a population (provided it is adjusted to the changing energy demands of the work environment), but it has less to say about changes in the consumption of services provided by the changing physical environment. This is particularly important during periods when the major cause of an increase in per capita income is the improvement in infrastructure, as it was in England from the eleventh to the eighteenth centuries.
47. Landes, *Prometheus*, p. 14.
48. For example, Jones, *Miracle*; and Mokyr, *Riches*.
49. Phillips, *Medieval Expansion*.
50. Snooks and McDonald, *Domesday Economy*; and Snooks, 'Rational calculation'. This theme is developed further in Chapter 6.
51. Kunitz, 'A long story', p. 275, brings these inadequate data together in tabular form.
52. For example, Gould, *Economic Growth*, and more recently Komlos, *Habsburg Monarchy*.
53. Maddison, *Capitalist Development*, p. 7.
54. North and Thomas, *Western World*.
55. Jones, 'Transactions costs'.
56. It is not hard to find contemporary examples of different costs of alternative economic and political systems. Eastern Europe has been grinding to a halt since the 1960s (see data in Maddison, 'European growth') under the inertia of economic systems that are unbelievably inefficient in the way resources, goods and services, and income are allocated.
57. See Snooks, 'Rational calculation', and Chapter 6 below.
58. The simulation employed in Figure 7.3 for the period 1086 to 1688 is based upon a number of solid data sets concerning the trend,

amplitude, and shape of the fluctuations in real GDP and GDP per capita. Once these key pieces of data are known we can, with reasonable confidence, simulate these macroeconomic variables. The procedure was as follows: (1) Real GDP was simulated for the period 1086 to 1688 and, as we know the population profile, GDP per capita was estimated as a residual. (2) The *trend* for real GDP was based upon the growth rate (0.49% p.a.) in Table 7.3. (3) The *shape* of the GDP curve was based upon the following data: (a) by combining the population series (interpolated between benchmark years) with the growth trend for GDP per capita (which grew by 0.29% p.a.) a damped outline of real GDP was obtained; (b) this outline was fleshed out by: (i) ensuring that the area between the simulated curve and the trend line, both above and below the line, is equalised; (ii) employing detailed historical evidence about the state of the economy from Hatcher, *English Economy*; Postan, *Medieval Economy*; Bolton, *English Economy*; Dyer, *Middle Ages*; Clarkson, *Pre-industrial Economy*; and (iii) adopting the assumption, discussed in the text, that there is a positive longrun relationship between population and real GDP. (4) The *amplitude* of the GDP per capita curve between 1300 and 1450 was limited to the increase in manorial income, deflated by the price of wheat, using data from Miller, *Ely*; Dyer, *Worcester*; and Gray, 'Incomes from land'.

59. The Winchester estate manors were scattered through the counties of Berkshire, Buckinghamshire, Hampshire, Oxfordshire, Somerset, Surrey and Wiltshire, together with the Isle of Wight. The number of observations on an annual basis is as high as forty. Yields did not begin to increase significantly until after 1500. See Titow, *English Rural Society*.
60. The Winchester yields are *average* yields for an estate that was not marginal. They cannot be used, therefore, as a test of the classical model.
61. Lee, 'Pre-industrial England', p. 606.
62. Hatcher, *English Economy*, pp. 69–70.
63. Hatcher, *English Economy*, pp. 33–4.
64. Abel, *Agricultural Fluctuations*, pp. 45 and 46.
65. Postan, *Medieval Economy*, pp. 275–6; see also p. 40.
66. Abel, *Agricultural Fluctuations*, p. 42.
67. Schofield, 'Wealth in England'.
68. Hatcher, *English Economy*, p. 33. See also Mokyr, *Riches*, chs 3 and 8.
69. Hatcher, *English Economy*, pp. 31–5, provides indirect evidence of a brief recovery.
70. I am currently investigating the feasibility of constructing an alternative set of wage data to that of Rogers over the last 800 years.
71. Clay, *Economic Expansion*, vol. I, p. 126, notes the slow increase in agricultural productivity in the sixteenth century. Also see Clarkson, *Pre-industrial Economy*, chs 3 to 5; and Wrigley, *Population*, p. 528.
72. Clay, *Economic Expansion*, p. 141.
73. The timing of the modern wave of economic changes presented here contradicts the interpretation of Wrigley, *Continuity*, and Persson,

Economic Growth, both of whom wish to push it forward into the early nineteenth century.

74. Lee, 'Pre-industrial England', p. 594.

8 TIME REGAINED?

1. I am currently working on a quantitative economic history of England, AD 1000–2000, which I also hope to extend to include Europe as a whole.
2. See Chapters 5 to 7 above.
3. An example of the latter attempt is that by Lee, 'Pre-industrial England'.
4. I am currently investigating the feasibility of collecting new raw data on wages in England.
5. See Snooks, 'What should economists be told about the past?'.

Bibliography

ABEL, W., *Agricultural Fluctuations in Europe: From the Thirteenth to the Twentieth Centuries* (London: Methuen, 1980).

AITKEN, H. G. J., 'On the present state of economic history', *Canadian Journal of Economic and Political Science*, 26 (February 1960) pp. 87–95.

ANDERSON, J. L. and JONES, E. L., 'Natural disasters and the historical response', *Australian Economic History Review*, 28 (March 1988) pp. 3–20.

ANDERSON, M., 'Household structure and the industrial revolution: middle nineteenth century Preston in comparative perspective', in P. Laslett and W. Wall (eds), *Household and Family in Past Time* (Cambridge University Press, 1972) pp. 215–35.

APPLEBY, J. O., *Economic Thought and Ideology in Seventeenth-century England* (Princeton, NJ: Princeton University Press, 1978).

ARNDT, H. W., 'A pioneer of national income estimates', *Economic Journal*, 59 (December 1949) pp. 616–25.

ARNDT, H. W., *A Course through Life: Memoirs of an Australian Economist* (Canberra: National Centre for Development Studies, Australian National University, 1985).

ARROW, K. J., 'The economic implications of learning by doing', *Review of Economic Studies*, 29 (June 1962) pp. 155–73.

ARROW, K. J., 'Economic theory and the hypothesis of rationality', in J. Eatwell, M. Milgate and P. Newman (eds), *The New Palgrave: A Dictionary of Economics*, vol. 2 (London: Macmillan, 1987) pp. 69–75.

ASHLEY, W. J., *An Introduction to English Economic History and Theory*, 2 vols (London: Longmans, 1913 and 1914; first published 1888 and 1893).

ASHLEY, W. J., 'On the study of economic history', in N. B. Harte (ed.), *The Study of Economic History: Collected Inaugural Lectures, 1893–1970* (London: Frank Cass, 1971) pp. 1–17.

ASTON, T. H., 'The origins of the manor in England', *Transactions of the Royal Historical Society* (1958) pp. 50–830.

AUERBACH, A. J., 'Taxation of income', in J. Eatwell, M. Milgate and P. Newman (eds), *The New Palgrave: A Dictionary of Economics*, vol. 4 (London: Macmillan, 1987) pp. 604–6.

AYMARD, M., 'Toward the history of nutrition: some methodological remarks', in R. Forster and O. Ranum (eds), *Food and Drink in History: Selections from Annales, Economies, Sociétés, Civilisations, vol. 5* (Baltimore, Md. and London: Johns Hopkins University Press, 1979) pp. 1–16.

BALLARD, A., *The Domesday Inquest* (London: Methuen, 1906).

BARNARD, A., *The Simple Fleece: Studies in the Australian Wool Industry* (Parkville: Melbourne University Press, 1962).

BARRO, R. J. and ROMER, P. M., 'Economic growth', *NBER Reporter* (Fall, 1990) pp. 1–5.

BECKER, G., *A Treatise on the Family* (Cambridge, Mass.: Harvard University Press, 1981).

BEDE, THE VENERABLE, *Ecclesiastical History of the English People* (Harmondsworth: Penguin, 1990).
BENHAM, F.C., *The Prosperity of Australia* (London: P.S. King and Son Ltd, 1928).
BEVERIDGE, W.H. *et al.*, *Prices and Wages in England from the Twelfth to the Nineteenth Century* (London: Longmans, Green, 1939).
BLAUG, M., *Economic Theory in Retrospect* (3rd edn) (Cambridge University Press, 1978).
BLAUG, M., *Great Economists before Keynes: An Introduction to the Lives and Works of One Hundred Economists of the Past* (Atlantic Highlands, NJ: Humanities Press International, 1986).
BOEHM, E.A., *Prosperity and Depression in Australia, 1887–1897* (Oxford University Press, 1971).
BOLLES, A.S., *Industrial History of the United States* (reprint of 3rd edn 1881) (New York: A.M. Kelley, 1966; first published 1878).
BOLTON, J.L., *The Medieval English Economy, 1150–1500* (London: Kent, 1980).
BOSWORTH, B.P., 'The global decline in saving: some international comparisons', in P. Stemp (ed.), *Saving and Policy* (Canberra: Australian National University, 1991) pp. 30–63.
BRAUDEL, F., *Civilization and Capitalism, 15th–18th Century*, vol. 1: *The Structures of Everyday Life: The Limits of the Possible* (London: Collins, 1981).
BRIDBURY, A.R., Review of Snooks and McDonald, *Domesday Economy*, *Economic Record*, 63 (December 1987) pp. 359–60.
BROWN, H.P. and HOPKINS, S.V. (eds), *A Perspective of Wages and Prices* (London and New York: Methuen, 1981).
BUTLIN, N.G., *Private Capital Formation in Australia: Estimates 1861–1900* (Canberra: Department of Economics, Australian National University, 1955).
BUTLIN, N.G., 'The shape of the Australian economy, 1861–1900', *Economic Record*, 47 (April 1958) pp. 10–29.
BUTLIN, N.G., *Australian Domestic Product, Investment and Foreign Borrowing: 1861–1938/39* (Cambridge University Press, 1962).
BUTLIN, N.G., *Investment in Australian Economic Development, 1861–1900* (Cambridge University Press, 1964).
BUTLIN, N.G. (ed.), *The Impact of Port Botany* (The Botany Bay Project Reports, No. 3) (Canberra: Australian National University Press, 1976).
BUTLIN, N.G., 'A fraternal farewell: tribute to S.J. Butlin', *Australian Economic History Review*, 18 (September 1978) pp. 99–118.
BUTLIN, N.G., *Our Original Aggression: Aboriginal Populations of Southeastern Australia, 1788–1850* (Sydney: George Allen & Unwin, 1983).
BUTLIN, N.G., *Economics and the Dreamtime* (Cambridge University Press, forthcoming).
BUTLIN, N.G. and ARNDT, H.W., *The Case for Bank Nationalisation* (Fabian Society Pamphlet No. 1) (Sydney: Fabian Society of NSW, 1947).
BUTLIN, N.G. and ARNDT, H.W., 'National output, income and expenditure of NSW, 1891', *Economic Record*, 26 (June 1950) pp. 30–49.

BUTLIN, N.G. and DE MEEL, H., *Public Capital Formation in Australia: Estimates 1860–1900* (Canberra: Department of Economics, Australian National University, 1954).

BUTLIN, N.G., BARNARD, A. and PINCUS, J.J., *Government and Capitalism: Public and Private Choice in Twentieth Century Australia* (Sydney: George Allen & Unwin, 1982).

BUTLIN, N.G., FITZGERALD, V.M. and SCOTT, R.H. (eds), *The Australian Economist, 1888–1898*, 2 vols (Sydney: Australian National University Press, 1986).

BUTLIN, N.G. and GREGORY, R.G., 'Trevor Winchester Swan, 1918–1989', *Economic Record*, 65 (December 1989) pp. 369–77.

BUTLIN, N.G. and SINCLAIR, W.A., 'Australian gross domestic product, 1788–1860', *Australian Economic History Review*, 26 (September 1986) pp. 96–125.

BUTLIN, S.J., 'An early estimate of Australian national income', *Economic Record*, 14 (December 1938) pp. 266–8.

BUTLIN, S.J., *Foundations of the Australian Monetary System: 1788–1951* (Melbourne University Press, 1953).

BUTLIN, S.J., *War Economy, 1939–1942* (Canberra: Australian War Memorial, 1955).

BUTLIN, S.J., *Australia and New Zealand Bank: The Bank of Australasia and the Union Bank of Australia Limited, 1828–1951* (London: Longmans, 1961).

BUTLIN, S.J., 'My own authority', *Australian Economic History Review*, 11 (March 1971) pp. 63–9.

BUTLIN, S.J., Reviews of John Hicks, *A Theory of Economic History* and W.H.B. Court, *Scarcity and Choice in History*, *Australian Economic History Review*, 11 (March 1971) pp. 77–9.

BUTLIN, S.J. and SCHEDVIN, C.B., *War Economy, 1942–1945* (Canberra: Australian War Memorial, 1977).

CAMPBELL, B.M.S., 'Agricultural progress in medieval England: some evidence from Eastern Norfolk', *Economic History Review*, 2nd series, 36 (1983) pp. 26–46.

CAMPBELL, B.M.S., 'Arable production in medieval England: some evidence from Norfolk', *Journal of Economic History*, 43 (1983) pp. 379–404.

CAMPBELL, B.M.S., 'The impact of concentrated urban demand upon agricultural production in late thirteenth-century England', Economic History Society Conference, April 1992.

CAPRIHAN, S.P., *Fight Against Hunger in Developing Countries* (Bhopal: J.K. Jain Brothers, 1975).

CARPENTER, K.J., 'The nutritive value of wheat proteins', in A. Spicer (ed.), *Bread: Social, Nutritional and Agricultural Aspects of Wheaten Bread* (London: Applied Science Publishers, 1975) pp. 93–114.

CAVE, M. and HARE, P., *Alternative Approaches to Economic Planning* (London: Macmillan, 1981).

CHANDAVARKAR, *IMF Staff Papers*, 24 (1977).

CIPOLLA, C., *Before the Industrial Revolution: European Society and Economy, 1000–1700* (London: Methuen, 1981).

CLANCHY, M. T., *England and its Rulers, 1066–1272* (Glasgow: Fontana, 1983).

CLAPHAM, J. H., 'Of empty economic boxes', *Economic Journal*, 32 (September 1922) pp. 305–14.

CLAPHAM, J.H., 'The economic boxes: a rejoinder', *Economic Journal*, 32 (December 1922) pp. 560–3.

CLAPHAM, J. H., *The Economic History of Modern Britain*, 3 vols (Cambridge University Press, 1926, 1932 and 1938).

CLAPHAM, J. H., *The Bank of England: A History*, 2 vols (Cambridge University Press, 1944).

CLAPHAM, J. H., *A Concise Economic History of Britain from the Earliest Times to 1750* (Cambridge University Press, 1949).

CLAPHAM, J. H., 'The study of economic history', in N. B. Harte (ed.), *The Study of Economic History* (London: Frank Cass, 1971) pp. 55–70.

CLARK, C. G., *The National Income, 1924–31* (London: Macmillan, 1932).

CLARK, C. G., *National Income and Outlay* (London: Macmillan, 1937).

CLARK, C. G., *The Conditions of Economic Progress* (London: Macmillan, 1940).

CLARK, C. G., 'Coghlan, Timothy (1855–1926)', in J. Eatwell, M. Milgate and P. Newman (eds), *The New Palgrave: A Dictionary of Economics*, vol. 1 (London: Macmillan, 1987) pp. 469–70.

CLARK, C. G. and CRAWFORD, J. G., *The National Income of Australia* (Sydney: Angus and Robertson, 1938).

CLARKSON, L. A., *The Pre-industrial Economy in England, 1500–1750* (London: Batsford, 1971).

CLAY, C. G. A., *Economic Expansion and Social Change: England 1500–1700*, 2 vols (Cambridge University Press, 1984).

CLIFFE LESLIE, T.E., *Essays in Political Economy* (2nd edn) (London: Longmans, Green & Co., 1888; reprinted New York: Augustus M. Kelley, 1969).

COATS, A. W., 'Disciplinary self-examination', *Scandinavian Economic History Review*, 38 (1990) pp. 3–18.

COGHLAN, T. A., *Wealth and Progress of New South Wales* (Sydney: Government Printer, 1887–1902).

COGHLAN, T. A., *A Statistical Account of the Seven Colonies of Australasia* (Sydney: Government Printer, 1890–1903/4).

COGHLAN, T. A., *Census of 1891 – Statistician's Report* (Sydney: Government Printer, 1891).

COGHLAN, T. A., *Labour and Industry in Australia, from the First Settlement in 1788 to the Establishment of the Commonwealth in 1901*, 4 vols (London and New York: Oxford University Press, 1918).

COLANDER, D. C. and COATS, A. W. (eds), *The Spread of Economic Ideas* (Cambridge and New York: Cambridge University Press, 1989).

COLANDER, D. C. and KLAMER, A., 'The making of an economist', *Journal of Economic Perspectives*, 1 (Fall, 1987) pp. 95–111.

COLE, H. S. D. *et al.* (eds), *Models of Doom: A Critique of The Limits to Growth* (New York: Universe Books, n.d. [1973]).

COLEMAN, D. C., *History and the Economic Past: An Account of the Rise and

Decline of Economic History in Britain (Oxford: Clarendon Press; and New York: Oxford University Press, 1987).
COLLINI, S. *et al.*, *That Noble Science of Politics: A Study in Nineteenth-Century Intellectual History* (Cambridge and New York: Cambridge University Press, 1983).
CONRAD, A.H. and MEYER, J.R., *The Economics of Slavery* (Chicago: Adline Publishing Company, 1964).
CORNWALL, J., 'Total factor productivity', in J. Eatwell, M. Milgate and P. Newman (eds), *The New Palgrave: A Dictionary of Economics*, vol. 4 (London: Macmillan, 1987) pp. 660–2.
CRAFTS, N.F.R., *British Economic Growth During the Industrial Revolution* (Oxford: Clarendon Press, 1985).
CUNNINGHAM, W., 'The perversion of economic history', *Economic Journal*, 2 (1892) pp. 491–506.
CUNNINGHAM, W., *The Growth of English Industry and Commerce*, 3 vols (5th edn) (Cambridge University Press, 1910 and 1912; first published 1882).
D'AMICO, R., *Historicism and Knowledge* (New York: Routledge, 1989).
DARBY, H.C., *Domesday England* (Cambridge University Press, 1979).
DARBY, H.C. and FINN, R.W. (eds), *The Domesday Geography of South-west England* (Cambridge University Press, 1967).
DARBY, H.C. and MAXWELL, I.S. (eds), *The Domesday Geography of Northern England* (Cambridge University Press, 1962).
DAVID, P., 'So, how would it matter if "history mattered"? Path-dependence in economics and its longrun implications', in G.D. Snooks (ed.), *Historical Analysis in Economics* (London: Routledge, 1993).
DE VRIES, C.A., 'Increasing crop yields – relative potential of specific crops by region and/or country', in Miloslav Rechcigl (ed.), *Man, Food, and Nutrition: Strategies and Technological Measures for Alleviating the World Food Problem* (Cleveland, Ohio: CRC Press, 1973) pp. 109–25.
DEANE, P., 'The implications of early national income estimates for the measurement of long-term economic growth in the United Kingdom', *Economic Development and Cultural Change*, 4 (1955) pp. 3–38.
DEANE, P., *The Evolution of Economic Ideas* (Cambridge and New York: Cambridge University Press, 1978).
DEANE, P., *The State and the Economic System: An Introduction to the History of Political Economy* (Oxford and New York: Oxford University Press, 1989).
DEATHERAGE, F.E., *Food for Life* (New York and London: Plenum Press, 1975).
DENISON, E.F., *Accounting for United States Economic Growth, 1927–1969* (Washington, DC: Brookings Institution, 1974).
DENISON, E.F., 'Growth accounting', in J. Eatwell, M. Milgate and P. Newman (eds), *The New Palgrave: A Dictionary of Economics*, vol. 2 (London: Macmillan, 1987) pp. 571–4.
DENNY, M., 'The relationship between functional forms for the production system', *Canadian Journal of Economics*, 7 (1974) pp. 21–31.
DOMAR, E.D., 'On the measurement of technological change', *Economic Journal*, 71 (December 1961) pp. 709–29.

DOMAR, E. D., 'Capital expansion rate of growth and employment', *Econometrica*, 14 (April 1964) pp. 137–47.
DOMAR, E. D., 'The causes of slavery or serfdom: a hypothesis', *Journal of Economic History*, 30 (1970) pp. 18–32.
DORFMAN, J., *Veblen and his America* (New York: The Viking Press, 1935).
DOUGLAS, P. H. and COBB, C. W., 'A theory of production', *American Economic Review*, Supplement 18 (March 1928) pp. 139–65.
DRUMMOND, I. M., *Political Economy at the University of Toronto* (Toronto: 1983).
DUNSDORFS, E., *The Australian Wheat Growing Industry, 1788–1948* (Melbourne University Press, 1956).
DYER, C., *Lords and Peasants in a Changing Society: The Estates of the Bishopric of Worcester, 680–1540* (Cambridge University Press, 1980).
DYER, C., 'English diet in the later Middle Ages', in T. H. Aston *et al.* (eds), *Social Relations and Ideas: Essays in Honour of R. H. Hilton* (Cambridge University Press, 1983) pp. 21–37.
DYER, C., 'Changes in diet in the late Middle Ages: the case of the harvest workers', *Agricultural History Review*, 36 (1988) pp. 21–37.
DYER, C., *Standards of Living in the Later Middle Ages: Social Change in England, 1200–1520* (Cambridge and New York: Cambridge University Press, 1989).
DYER, C., 'The consumer and the market in the later Middle Ages', *Economic History Review*, 2nd series, 42 (August 1989) pp. 305–27.
EATWELL, J., MILGATE, M. and NEWMAN, P. (eds), *The New Palgrave: A Dictionary of Economics*, 4 vols (London: Macmillan, 1987).
EHRLICH, I., 'The problem of development: introduction', *Journal of Political Economy*, 98 (October 1990) pp. S1–S11.
EHRLICH, P. R., *The Population Bomb* (New York: Ballantine Books, 1970).
EHRLICH, P. R. and EHRLICH, A. H., *Population, Resources, Environment: Issues in Human Ecology* (San Francisco: W. H. Freeman, 1972).
EHRLICH, P. R. and EHRLICH, A. H., *The End of Affluence: A Blueprint for Your Future* (New York: Ballantine Books, 1975).
EHRLICH, P. R. and EHRLICH, A. H., *The Population Explosion* (New York: Simon and Schuster, n.d. [c. 1990]).
EHRLICH, P. R., EHRLICH, A. H. and HOLDREN, J. P., *Human Ecology: Problems and Solutions* (San Francisco: W. H. Freeman, 1973).
EHRLICH, P. R. and HARRIMAN, R. L., *How to be a Survivor* (New York: Ballantine Books, 1975).
EHRLICH, P.R. and HOLDREN, J. P. (eds), *The Cassandra Conference: Resources and the Human Predicament* (College Station, Tex.: Texas A & M University Press, 1968).
EHRLICH, P. R. and PIRAGES, D. C., *Ark II: Social Response to Environmental Imperatives* (San Francisco: W. H. Freeman, 1974).
EISNER, R., 'Extended accounts for national income and product', *Journal of Economic Literature*, 26 (December 1988) pp. 1611–84.
ELIOT, T. S. 'East Coker', *Collected Poems 1902–62* (London: Faber & Faber, 1963) pp. 196–204.
ELLENBY, J., *The Anglo-Saxon Household* (Cambridge University Press, 1989).

ELLIOTT, R. W. V., *Runes: An Introduction* (Manchester University Press, 1959).

ESTRIN, S. and HOLMES, P., *French Planning in Theory and Practice* (London and Boston, Mass.: George Allen & Unwin, 1983).

EYTON, R.W., *Key to Domesday: An Analysis and Digest of the Dorset Survey* (London and Dorchester: Taylor, 1878).

FARBER, H, 'A price and wage study for northern Babylonia during the Old Babylonia period', *Journal of Economic and Social History of the Orient*, 21 (1978) pp. 1–51.

FARMER, D. L., 'Some price fluctuations in Angevin England', *Economic History Review*, 2nd series, 9 (August 1956) pp. 34–43.

FENOALTEA, S., 'Authority, efficiency and agricultural organization in medieval England and beyond: a hypothesis', *Journal of Economic History*, 35 (1975) pp. 693–718.

FENOALTEA, S., 'Risk, transaction costs, and the organization of medieval agriculture', *Explorations in Economic History*, 13 (1976) pp. 129–51.

FIELD, A. J., 'The future of economic history', in A. J. Field (ed.), *The Future of Economic History* (Boston, Mass.: Kluwer Nijhoff Publishing, 1987) pp. 1–41.

FINLEY, M. I., *The Ancient Economy* (Berkeley, Calif.: University of California Press, 1973).

FLACELIERE, R., *Greek Oracles* (London: Elek Books, 1965).

FOGEL, R. W., *Railroads and American Economic Growth: Essays in Econometric History* (Baltimore, Md: Johns Hopkins University Press, 1964).

FOGEL, R. W., 'Nutrition and the decline in mortality since 1700: some additional preliminary findings', Working Paper No. 1802, *NBER* (1986).

FOGEL, R. W. and ENGERMAN, S. L., *Time on the Cross*, 2 vols (Boston, Mass.: Little, Brown, 1974).

FONTENROSE, J., *The Delphic Oracle: Its Responses and Operations with a Catalogue of Responses* (Berkeley, Calif.: University of California Press, n.d. [c. 1978]).

FORSTER, C., *Industrial Development in Australia, 1920–30* (Canberra: Australian National Univerity Press, 1964).

FOXWELL, H. S., 'The economic movement in England', *Quarterly Journal of Economics*, 2 (October 1887) pp. 84–103.

GALBRAITH, V. H., *The Making of Domesday Book* (Oxford: Clarendon Press, 1961).

GODFREY, L. G. and WICKENS, M. R., 'Testing linear and log-linear regressions for functional form', *Review of Economic Studies*, 48 (1981) pp. 487–96.

GOLDSMITH, R.W., 'An estimate of the size and structure of the national product of the early Roman Empire', *Review of Income and Wealth*, Series 30 (September 1984) pp. 263–88.

GORDON, A. and SUZUKI, D., *It's a Matter of Survival* (North Sydney: George Allen & Unwin, 1990).

GOULD, J. D., *Economic Growth in History: Survey and Analysis* (London: Methuen, 1972).

GRAS, N. S. B., 'The present condition of economic history', *The Quarterly Journal of Economics*, 34 (February 1920) pp. 209–24.

GRAS, N. S. B., 'The rise and development of economic history', *Economic History Review*, 1 (January 1927) pp. 12–34.
GRAS, N. S. B., 'Economic history in the United States', in E. R. A. Seligman (ed.), *The Encyclopedia of the Social Sciences*, vol. 5 (New York: Macmillan, 1931) pp. 325–7.
GRAY, H. L., 'Incomes from land in England in 1436', *English Historical Review*, 49 (October 1934) pp. 607–39.
GREEN, J. A., 'The last century of Danegeld', *English Historical Review*, 96 (1981) pp. 241–58.
GRILICHES, Z., 'Productivity: measurement problems', in J. Eatwell, M. Milgate and P. Newman (eds), *The New Palgrave: A Dictionary of Economics*, vol. 3 (London: Macmillan, 1987) pp. 1010–13.
GROENEWEGEN, P. D., '"Political economy" and "economics"', in J. Eatwell, M. Milgate and P. Newman (eds), *The New Palgrave: A Dictionary of Economics*, vol. 3 (London: Macmillan, 1987) pp. 904–07.
GROENEWEGEN, P. D. and McFARLANE, B., *A History of Australian Economic Thought* (London and New York: Routledge, 1990).
GROSSMAN, G. M. and HELPMAN, E., 'Trade, innovation and growth', *American Economic Review*, 80 (May 1990) pp. 86–91.
HAHN, F. H., 'Neoclassical growth theory', in J. Eatwell, M. Milgate and P. Newman (eds), *The New Palgrave: A Dictionary of Economics*, vol. 3 (London: Macmillan, 1987) pp. 625–34.
HAHN, F. H. and MATTHEWS, R. C. D., 'The theory of economic growth: a survey', *Economic Journal*, 74 (December 1964) pp. 779–902.
HAIG, B. D., *Capital Stock in Australian Manufacturing* (Canberra: Australian National University, 1980).
HAMILTON, N.E. S. A. (ed.), *Inquisitio Comitatus Cantabrigiensis* (London: Royal Society of Literature, 1876).
HANSEN, W. L., 'Educating and training new economics Ph.D.s: how good a job are we doing?' *American Economic Review*, 80 (May 1990) pp. 437–44.
HARRIS, D. J., 'Classical growth models', in J. Eatwell, M. Milgate and P. Newman (eds), *The New Palgrave: A Dictionary of Economics*, vol. 1 (London: Macmillan, 1987) pp. 445–9.
HARROD, R. F., 'An essay in dynamic theory', *Economic Journal*, 49 (March 1939) pp. 14–33.
HARTE, N. B. (ed.), *The Study of Economic History: Collected Inaugural Lectures, 1893–1970* (London: Frank Cass, 1971).
HATCHER, J., *Plague, Population, and the English Economy, 1348–1530* (London: Macmillan, 1977).
HAWKE, G. R., *Railways and Economic Growth in England and Wales, 1840–1870* (Oxford: Clarendon Press, 1970).
HAWKE, G. R., *The Making of New Zealand: An Economic History* (Cambridge and New York: Cambridge University Press, 1985).
HAYEK, F. A., 'The London School of Economics, 1895–1945', *Economica*, New Series 13 (February 1946) pp. 1–31.
HICKS, N. D., *This Sin and Scandal: Australia's Population Debate, 1891–1911* (Canberra: Australian National University Press, 1978).
HIGGINS, B., *Economic Development: Principles, Problems and Policies* (London: Constable, 1959).

HILTON, R. H., *A Medieval Society: The West Midlands at the End of the Thirteenth Century* (Cambridge University Press, 1966).

HOLLISTER, C. W., *The Military Organisation of Norman England* (Oxford: Clarendon Press, 1965).

HOLMES, P., 'Indicative planning', in J. Eatwell, M. Milgate and P. Newman (eds), *The New Palgrave: A Dictionary of Economics*, vol. 2 (London: Macmillan, 1987) pp. 781–3.

HOLMES, R., *Coleridge: Early Visions* (London: Hodder & Stoughton, 1989).

INADA, K., 'On a two-sector model of economic growth: comments and a generalisation', *Review of Economic Studies*, 30 (1963) pp. 119–27.

INDECS, *State of Play 6* (Sydney: George Allen & Unwin, 1990).

INGRAM, J. K., 'Address of the president of section F of the British Association, at the forty-eighth meeting, held at Dublin, in August 1878', *Journal of the Royal Statistical Society* (December, 1878) pp. 602–29.

INGRAM, J. K., *A History of Political Economy* (London: Black, 1915; first published 1888).

JACKSON, R.V., 'Growth and deceleration in English agriculture, 1660–1790', *Economic History Review*, 2nd series, 38 (August 1985) pp. 333–51.

JOHNS, C. H. W., *An Assyrian Doomsday Book or Liber Censualis of the District Round Harran in the Seventh Century B.C.* (Leipzig: J. C. Hinrichs'sche Buchhandlung, 1901).

JONES, E. L., *Agriculture and Economic Growth in England, 1650–1815* (London: Methuen, 1967).

JONES, E. L., *The European Miracle: Environments, Economies and Geopolitics in the History of Europe and Asia* (Cambridge University Press, 1981).

JONES, E. L., *Growth Recurring: Economic Change in World History* (Oxford: Clarendon Press, 1988).

JONES, E. L., 'Recurrent transitions to *intensive* growth', in J. Goudsblom *et al.* (eds), *Human History and Social Progress* (University of Exeter Press, 1989) pp. 46–62.

JONES, R., *An Essay on the Distribution of Wealth and on the Sources of Taxation*, vol. 1: *Rent* (New York: Augustus M. Kelley, 1964; first published 1831).

JONES, S. R. H., Transactions costs, institutional change, and the emergence of a market economy in later Anglo-Saxon England (unpublished paper, Auckland, 1989).

KADISH, A., *Historians, Economists, and Economic History* (London and New York: Routledge, 1989).

KALDOR, N., 'Alternative theories of distribution', *Economic Journal*, 23 (1955) pp. 83–100.

KENDRICK, J. W., *Postwar Productivity Trends in the United States, 1948–1969* (New York: National Bureau of Economic Research and Columbia University Press, 1973).

KENWOOD, A. G. and LOUGHEED, A. L., *The Growth of the International Economy, 1820–1980: An Introductory Text* (London: George Allen & Unwin, 1983).

KEYNES, J. M., *The General Theory of Employment, Interest, and Money* (London: Macmillan, 1936).

KEYNES, J. M., 'William Stanley Jevons, 1835–1882: a centenary allocution on his life and work as economist and statistician', *Journal of Royal Statistical Society*, New Series, 99 (1936) pp. 516–48.
KEYNES, J. M., *A Treatise on Probability* (London: Macmillan, 1948).
KEYNES, J. N., *The Scope and Method of Political Economy* (London: Macmillan, 1891).
KING, G., *Two Tracts* (edited by G.E. Barnett) (Baltimore, Md: Johns Hopkins Press, 1936).
KOMLOS, J., *Nutrition and Economic Development in the Eighteenth-Century Habsburg Monarchy: An Anthropometric History* (Princeton, NJ: Princeton University Press, 1989).
KOOT, G. M., *English Historical Economics, 1870–1926: The Rise of Economic History and Neomercantilism* (Cambridge University Press, 1987).
KUNITZ, S. J., 'Making a long story short: a note on men's height and mortality in England from the first through the nineteenth centuries', *Medical History*, 31 (1987) pp. 269–80.
KUZNETS, S., *National Income and Capital Formation, 1919–1935* (New York: National Bureau of Economic Research, 1937).
KUZNETS, S., *Commodity Flow and Capital Formation* (New York: National Bureau of Economic History, 1938).
KUZNETS, S., *Modern Economic Growth: Rate, Structure, and Spread* (New Haven, Conn.: Yale University Press, 1966).
LA NAUZE, J. A., *Political Economy in Australia: Historical Studies* (Melbourne University Press, 1949).
LANDES, D. S., *The Unbound Prometheus: Technological Change and Industrial Development in Western Europe from 1750 to the Present* (London: Cambridge University Press, 1969).
LEAVER, R. A., 'Five-hides in ten counties: a contribution to the Domesday regression debate', *Economic History Review*, 2nd series, 16 (1988) pp. 525–42.
LEE, R., 'Population in pre-industrial England: an econometric analysis', *Quarterly Journal of Economics*, 87 (1973) pp. 581–607.
LENNARD, R., *Rural England, 1086–1135: A Study of Social and Agrarian Conditions* (Oxford: Clarendon Press, 1959).
LEONTIEF, W. W., *Input–Output Economics* (New York: Oxford University Press, 1966).
LEONTIEF, W. *et al.*, *Studies in the Structure of the American Economy: Theoretical and Empirical Explorations in Input–Output Analysis* (New York: Oxford University Press, 1953).
LEVI, M., *Of Rule and Revenue* (Berkeley, Calif.: University of California Press, 1988).
LINDERT, P. H., 'English occupations: 1670–1811', *Journal of Economic History*, 40 (1980) pp. 685–712.
LINDERT, P. H. and WILLIAMSON, J. G., 'Revising England's social tables, 1688–1812', *Explorations in Economic History*, 19 (1982) pp. 385–408.
LLOYD, T. H., 'The movement of wool prices in medieval England', *Economic History Review*, Supplement No. 6 (1973).
LOYN, H. R., *The Vikings in Britain* (London: Batsford, 1977).

LOYN, H. R., *Anglo-Saxon England and the Norman Conquest* (London: Longmans, 1962).
LUCAS, R.E. Jr, 'On the mechanics of economic development', *Journal of Monetary Economics*, 22 (January 1988) pp. 3–22.
MADDISON, A., 'Per capita output in the long-run', *Kyklos*, 32 (1979) pp. 412–29.
MADDISON, A., *Phases of Capitalist Development* (New York: Oxford University Press, 1982).
MADDISON, A., Measuring European growth: the core and the periphery (paper presented to the Tenth International Conference in Economic History, Leuven, Belgium, 1990).
MAITLAND, F. W., *Domesday Book and Beyond: Three Essays in the Early History of England* (Cambridge University Press, 1921).
MALONEY, J. *Marshall, Orthodoxy and the Professionalisation of Economics* (Cambridge University Press, 1985).
MALTHUS, T. R., *An Essay on the Principle of Population as it Affects the Future Improvement of Society*, 2 vols (revised edn) (London: Dent, 1803; first published 1798).
MARSHALL, A., 'A reply', *Economic Journal*, 2 (1892) pp. 507–19.
MARSHALL, A., *Principles of Economics* (4th edn) (London: Macmillan, 1898).
MATTHEWS, R.C.O., FEINSTEIN, C. H. and ODLING-SMEE, J.C., *British Economic Growth, 1856–1973* (Stanford, Calif.: Stanford University Press, 1982).
MATTHEWS, R. C. O. and SUPPLE, B., The ordeal of economic freedom: Marshall on economic history (unpublished paper).
McCLOSKEY, D. N. (ed.), *Essays on a Mature Economy: Britain after 1840* (London: Methuen, 1971).
McCLOSKEY, D. N., *Economic Maturity and Entrepreneurial Decline: British Iron and Steel, 1870–1913* (Cambridge, Mass.: Harvard University Press, 1973).
McCLOSKEY, D. N., 'English open fields as behaviour towards risk', *Research in Economic History*, 1 (1976) pp. 124–70.
McCLOSKEY, D. N., *Econometric History* (London: Macmillan, 1987).
McCLOSKEY, D. N., 'The rhetoric of statistics', in J. Eatwell, M. Milgate and P. Newman (eds), *The New Palgrave: A Dictionary of Economics*, vol. 1 (London: Macmillan, 1987) pp. 173–4.
McCLOSKEY, D. N. and NASH, J., 'Corn at interest: the cost of grain storage in medieval England', *American Economic Review*, 24 (March 1984) pp. 174–87.
McLEAN, I. W., 'The adoption of harvest machinery in Victoria in the late nineteenth century', *Australian Economic History Review*, 13 (March 1973) pp. 41–56.
McLEAN, I. W. and MADDOCK, R. (eds), *The Australian Economy in the Long Run* (Cambridge University Press, 1987).
MEADE, J.E., *The Theory of Indicative Planning* (Manchester University Press, 1970).
MEADE, J.E. and HAHN, F. H., 'The rate of profit in a growing economy', *Economic Journal*, 75 (June 1966) pp. 445–8.

MEADOWS, D. H. *et al., The Limits to Growth: A Report for the Club of Rome's Project on the Predicaments of Mankind* (New York: Universe Books, 1972).

MEIER, G. M., *Leading Issues in Economic Development* (New York: Oxford University Press, 1989).

MERRETT, D. T., 'Banking practice', *Australian Economic History Review*, 29 (March 1989) pp. 60–85.

MILLER, E., *The Abbey and Bishopric of Ely: The Social History of an Eccelesiastical Estate from the Tenth Century to the Early Fourteenth Century* (Cambridge University Press, 1951).

MILLER, E. and HATCHER, J., *Medieval England: Rural Society and Economic Change, 1086–1348* (London: Longmans, 1978).

MODIGLIANI, F. and SAMUELSON, P. A., 'The Pasinetti Paradox in neo-classical and more general models', *Review of Economic Studies*, 33 (1966) pp. 269–301.

MOKYR, J., *The Lever of Riches: Technological Creativity and Economic Progress* (New York: Oxford University Press, 1990).

MOKYR, J., Evolutionary biology, technological change, and economic history (seminar paper presented to the Economic History Joint Seminar Program, Australian National University, Canberra, 23 July 1990).

MORGAN, M. S., *The History of Econometric Ideas* (Cambridge and New York: Cambridge University Press, 1990).

MORGAN, M. S., The stamping out of process analysis in econometrics (unpublished paper, Department of Economic History, London School of Economics).

MORISHIMA, M., *Equilibrium Stability and Growth* (Oxford: Clarendon Press, 1964).

MULHALL, M. G., *The Dictionary of Statistics* (London: George Routledge & Sons, 1884).

NORTH, D. and THOMAS, R. P., 'The rise and fall of the manorial system: a theoretical model', *Journal of Economic History*, 31 (1971) pp. 777–803.

NORTH, D. and THOMAS, R. P., *The Rise of the Western World* (Cambridge University Press, 1973).

OFFER, A., 'Using the past in Britain: retrospect and prospect', *The Public Historian*, 6 (1984).

OHKAWA, K. and ROSOVSKY, H., *Japanese Economic Growth: Trend Acceleration in the Twentieth Century* (Stanford, Calif.: Stanford University Press, 1973).

PAGAN, A. R. and HALL, A. D., 'Diagnostic tests as residual analysis', *Econometric Reviews*, 2 (1983) pp. 159–218.

PAINTER, S., *Studies in the History of the English Feudal Barony* (Baltimore, Md: Johns Hopkins University Press, 1943).

PALGRAVE, R. H. I. (ed.), *Dictionary of Political Economy*, 3 vols (London: Macmillan & Co., 1894, 1896 and 1899).

PASINETTI, L. L., 'Rate of profit and income distribution in relation to the rate of economic growth', *Review of Economic Studies*, 29 (1962) pp. 267–79.

PAUL, A. A. and SOUTHGATE, D. A. T. (eds), *McCance and Widdowson's The Composition of Foods* (London: HMSO, 1978).

PEARSON, C.H., *History of England during the Early and Middle Ages* (London: Bell & Daldy, 1867).
PERSSON, K.G., *Pre-industrial Economic Growth* (Oxford: Basil Blackwell, 1988).
PERSSON, K.G., 'Labour productivity in medieval agriculture: Tuscany and the "Low Countries"', in B.M.S. Campbell and M. Overton (eds), *Land, Labour and Livestock: Historical Studies in European Agricultural Productivity* (Manchester and New York: Manchester University Press, 1991) pp. 124–45.
PHILLIPS, J.R.S., *The Medieval Expansion of Europe* (Oxford: Clarendon Press, 1988).
PIGOU, A.C., 'Empty economic boxes: a reply', *Economic Journal*, 32 (December 1922) pp. 458–65.
PIGOU, A.C., 'The laws of diminishing and increasing cost', *Economic Journal*, 37 (June 1927) pp. 188–97.
PIGOU, A.C., 'Wage policy and unemployment', *Economic Journal*, 37 (September 1927) pp. 355–68.
PIGOU, A.C., 'An analysis of supply', *Economic Journal*, 38 (June 1928) pp. 238–57.
PIGOU, A.C., 'The statistical derivation of demand curves', *Economic Journal*, 40 (September 1970) pp. 384–400.
POLLARD, S. and CROSSLEY, D.W., *The Wealth of Britain, 1085–1966* (London: Batsford, 1968).
POOLE, R.L., *The Exchequer in the Twelfth Century* (Oxford: Clarendon Press, 1912).
POPE, D.H., 'Australian immigration: a critique of the push–pull model', *Australian Economic History Review*, 16 (September 1976) pp. 375–421.
POSTAN, M.M., 'Medieval agrarian society in its prime: England', in M.M. Postan (ed.), *The Cambridge Economic History of Europe*, vol. 1: *The Agrarian Life of the Middle Ages* (2nd edn) (Cambridge University Press, 1966) pp. 549–632.
POSTAN, M.M., *The Medieval Economy and Society: An Economic History of Britain in the Middle Ages* (London: Weidenfeld & Nicolson, 1972).
POSTAN, M.M., *Essays on Medieval Agriculture and General Problems of the Medieval Economy* (London: Cambridge University Press, 1973).
POSTAN, M.M., 'Feudalism and its decline: a semantic exercise', in T.H. Aston *et al.* (eds), *Social Relations and Ideas: Essays in Honour of R.H. Hilton* (Cambridge University Press, 1983) pp. 73–87.
POSTGATE, J.N., *Taxation and Consumption in the Assyrian Empire* (Rome: Biblical Institute Press, 1974).
PYKE, M., *Man and Food* (London: Weidenfeld & Nicolson, 1970).
RAZI, Z., *Life, Marriage and Death in a Medieval Parish: Economy, Society and Demography in Halesowen, 1270–1400* (Cambridge University Press, 1980).
ROBBINS, L., 'The representative firm', *Economic Journal*, 38 (September 1928) pp. 387–404.
ROBERTSON, D.H., 'Those empty boxes', *Economic Journal*, 34 (March 1924) pp. 16–30.

ROBERTSON, D. H., SHOVE, G. F. and SRAFFA, P., 'Increasing returns and the representative firm: a symposium', *Economic Journal*, 40 (March 1930) pp. 79–116.

ROBINSON, J. V., *Collected Economic Papers* (Oxford: Basil Blackwell, 1965).

ROGERS, J. E. T., *A History of Agriculture and Prices in England: From the Year after the Oxford Parliament (1259) to the Commencement of the Continental War (1793)*, 7 vols (Oxford: Clarendon Press, 1866–1902).

ROMER, P. M., 'Increasing returns and long-run growth', *Journal of Political Economy*, 94 (1986) pp. 1002–37.

ROSTOW, W. W., *The Stages of Economic Growth: A Non-Communist Manifesto* (Cambridge University Press, 1960).

ROUND, J. H., *Feudal England: Historical Studies on the Eleventh and Twelfth Centuries* (London: Swan Sonnenschein, 1895).

RUSSELL, J. C., *British Medieval Population* (Albuquerque, N. Mex.: University of New Mexico Press, 1948).

SAMUELSON, P. A., 'Economists and the history of ideas', *American Economic Review*, 52 (1962) pp. 1–18.

SATO, K., 'A two-level constant elasticity of substitution production function', *Review of Economic Studies*, 34 (1967) pp. 201–18.

SCHEDVIN, C. B., *Australia and the Great Depression. A study of Economic Development and Policy in the 1920s and 1930s* (Sydney University Press, 1970).

SCHEDVIN, C. B., *Shaping Science and Industry: A History of Australia's Council for Scientific and Industrial Research, 1926–49* (Sydney: George Allen & Unwin, 1987).

SCHOFIELD, R. S., 'The geographical distribution of wealth in England, 1334–1649', *Economic History Review*, 2nd series, 18 (December 1965) pp. 483–510.

SCHOFIELD, S., *Development and the Problems of Village Nutrition* (London: Croom Helm, 1979).

SCHUMPETER, J. A., 'The instability of capitalism', *Economic Journal*, 38 (September 1928) pp. 361–86.

SCHUMPETER, J. A., *The Theory of Economic Development: An Inquiry into Profits, Capital, Credit, Interest and the Business Cycle* (Cambridge, Mass.: Harvard University Press, 1949).

SCHUMPETER, J. A., *History of Economic Analysis* (edited by E. B. Schumpeter) (New York: Oxford University Press, 1954).

SELIGMAN, E., *Essays in Taxation* (10th edn) (New York: John Wiley, 1957).

SHANN, E. O. G., *Cattle Chosen: The Story of the First Group Settlement in Western Australia* (London: Oxford University Press, 1926).

SHANN, E. O. G., *The Boom of 1890 and Now: A call to Australia to put her house in order lest drought and falling prices for wool and wheat overtake us* (Sydney: Cornstalk, 1927).

SHANN, E. O. G., *Bond or Free?* (Sydney: Angus and Robertson, 1930).

SHANN, E. O. G., *An Economic History of Australia* (Cambridge University Press, 1930).

SHEFFIELD, B., *Observations on the Commerce of the American States* (London: J. Debrett, 1784).

SHOVE, G. H., 'Varying costs and marginal net products', *Economic Journal*, 38 (June 1928) pp. 258–66.
SIMON, H. A., *Models of Man* (New York: John Wiley, 1957).
SIMON, H. A., *Models of Bounded Rationality*, 2 vols (Cambridge, Mass.: MIT Press, 1982).
SIMON, H. A., 'Bounded rationality', in J. Eatwell, M. Milgate and P.Newman (eds), *The New Palgrave: A Dictionary of Economics*, vol. 1 (London: Macmillan, 1987) pp. 266–7.
SINCLAIR, W. A., *The Process of Economic Development in Australia* (Melbourne: Cheshire, 1976).
SINCLAIR, W. A., 'Gross domestic product', in W. Vamplew *et al.* (eds), *South Australian Historical Statistics* (Sydney: University of New South Wales, 1988) pp. 25–36.
SMITH, R. M., 'Demographic developments in rural England, 1300–1348: a survey', in B. M. S. Campbell (ed.), *Before the Black Death: Studies in the 'Crisis' of the Early Fourteenth Century* (Manchester and New York: Manchester University Press, 1991) pp. 25–77.
SMITH, V. L., 'Experimental methods in econometrics', in J. Eatwell, M. Milgate and P. Newman (eds), *The New Palgrave: A Dictionary of Economics*, vol. 2 (London: Macmillan, 1987) pp. 241–9.
SNOOKS, G. D., 'Regional estimates of GDP', *Economic Record*, 48 (December 1972) pp. 536–53.
SNOOKS, G. D., 'Innovation and the growth of the firm: Hume Enterprises 1910–40', *Australian Economic History Review*, 13 (March 1973) pp. 16–40.
SNOOKS, G. D., 'The growth process of the firm: a case study', *Australian Economic Papers*, 12 (December 1973) pp. 162–74.
SNOOKS, G. D., *Depression and Recovery in Western Australia, 1928/29–1938/39: A Study in Cyclical and Structural Change* (Nedlands, WA: University of Western Australia, 1974).
SNOOKS, G. D., 'Orthodox and radical interpretations of the development of Australian capitalism', *Labour History*, 28 (May 1975) pp. 1–11 and 18–21.
SNOOKS, G. D., 'A note on the use of alternative total factor productivity indexes', *Economic Record*, 52 (September 1976) pp. 373–7.
SNOOKS, G. D., 'The arithmetic of regional growth', *Australian Economic History Review*, 19 (March 1979) pp. 63–74.
SNOOKS, G. D., 'Growth and productivity change in the Australian mechanical engineering industry, 1910–1940', *Australian Economic History Review*, 24 (March 1984) pp. 53–70.
SNOOKS, G. D., 'Shann, Edward Owen Giblin (1884–1935)', *Australian Dictionary of Biography*, vol. 11 (Melbourne University Press, 1988) pp. 574–6.
SNOOKS, G. D., 'Estimating Australian household labour services, 1881–1986', *Working Papers in Economic History*, No. 130 (Australian National University, December 1989).
SNOOKS, G. D., 'Economic growth during the last millennium: a quantitative perspective for the British Industrial Revolution', *Working Papers in Economic History*, No. 140 (Australian National University, July 1990).

SNOOKS, G. D., 'Arbitrary decree or rational calculation? The contribution of Domesday Book to economic history and economics', *Australian Economic History Review*, 30 (September 1990) pp. 23–49.
SNOOKS, G. D., 'What should economists be told about the past? A review article', *Australian Economic History Review*, 30 (September 1990) pp. 89–94.
SNOOKS, G. D., '"In my beginning is my end": the life and work of Noel George Butlin, 1921–1991', *Australian Economic History Review*, 31 (September 1991) pp. 3–27.
SNOOKS, G. D., 'It's time to take time into account: economic history in the 1990s and beyond' (1990 Australian Historical Association/Westpac Lecture), *Working Papers in Economic History*, No. 143 (Australian National University, October 1990).
SNOOKS, G. D., *Economic Portrait of the Family: The Household Economy in Australia, 1788–1990* (Cambridge University Press, 1993).
SNOOKS, G. D. (ed.), *Historical Analysis in Economics* (London: Routledge, 1993).
SNOOKS, G. D., *The Doomsday Debate: Economist Versus Environmentalist* (forthcoming).
SNOOKS, G. D. and McDONALD, J., 'The determinants of manorial income in Domesday England: evidence from Essex', *Journal of Economic History*, 45 (1985) pp. 541–56.
SNOOKS, G.D. and McDONALD, J., 'How artificial were the tax assessments of Domesday England? The case of Essex', *Economic History Review*, 2nd series, 38 (1985) pp. 352–72.
SNOOKS, G. D. and McDONALD, J., 'Statistical analysis of Domesday Book (1086)', *Journal of the Royal Statistical Society*, Series A, 148 (1985) pp. 147–60.
SNOOKS, G. D. and McDONALD, J., *Domesday Economy: A New Approach to Anglo-Norman History* (Oxford: Clarendon Press, 1986).
SNOOKS, G. D. and McDONALD, J., 'The suitability of Domesday Book for cliometric analysis', *Economic History Review*, 2nd series, 40 (May 1987) pp. 252–61.
SNOOKS, G. D. and McDONALD, J., 'A case of mistaken identity: national taxation and local authority administration in Domesday England', *Working Papers in Economic History*, No. 139 (Australian National University, June 1990).
SOLOW, R. M., 'A contribution to the theory of economic growth', *Quarterly Journal of Economics*, 70 (February 1956) pp. 65–94.
SOLOW, R. M., 'Technological change and the aggregate production function', *Review of Economics and Statistics*, 39 (August 1957) pp. 312–20.
SOLOW, R. M., 'Substitution and fixed proportions in the theory of capital', *Review of Economic Studies*, 29 (1962) pp. 207–18.
SOLOW, R. M., 'Faith, hope, and clarity', in D. C. Colander and A. W. Coats (eds), *The Spread of Economic Ideas* (Cambridge and New York: Cambridge University Press, 1989) pp. 37–41.
SRAFFA, P., 'The laws of returns under competitive conditions', *Economic Journal*, 36 (December 1926) pp. 535–50.

STENTON, F. M., *Anglo-Saxon England* (Oxford: Clarendon Press, 1943).
STEPHENSON, C., *Medieval Institutions: Selected Essays* (edited by B. D. Lyon) (Ithaca, NY: Cornell University Press, 1954).
STIGLER, G. J., 'Palgrave's dictionary of economics', *Journal of Economic Literature*, 26 (December 1988) pp. 1729–36.
STONE, J. O., '1929 and all that', *Quadrant*, 28 (October 1984) pp. 9–20.
STONE, J. R. N., 'The accounts of society: Nobel Memorial Lecture', *Les Prix Nobel en 1984* (Stockholm: 1984).
STONE, R., 'Some seventeenth century econometrics: consumers' behaviour', *Revue européenne des sciences sociales*, 26 (1988) pp. 19–41.
STONE, R., 'When will the war end?', *Cambridge Journal of Economics*, 12 (1988) pp. 193–201.
STONE, R., 'Some seventeenth century econometrics: public finance', *Revue européenne des sciences sociales*, 26 (1989) pp. 5–32.
SUTCLIFFE, J. T., *The National Dividend* (Melbourne University Press, 1926).
SWAN, T. W., 'Economic growth and capital accumulation', *Economic Record*, 32 (November 1956) pp. 334–61.
SYME, D., 'On the method of political economy', *Westminster Review* (July 1871) pp. 206–18.
SYME, D., *Outlines of an Industrial Science* (London: Henry S. King & Co., 1876).
TAKAYAMA, A., 'On a two-sector model of economic growth: a comparative statics analysis', *Review of Economic Studies*, 36 (1963) pp. 95–104.
TAYLOR, T. G., *Principles of Human Nutrition* (London: Edward Arnold, 1978).
TITOW, J. Z., *English Rural Society, 1200–1350* (London: George Allen & Unwin, 1969).
TITOW, J. Z., Review of G. D. Snooks and J. McDonald, *Domesday Economy*, *Economic History Review*, 2nd series, 41 (May 1988) pp. 302–3.
TOYNBEE, A., *The Industrial Revolution* (Boston, Mass.: Beacon Press, 1962).
UNITED NATIONS, *Scope and Structure of the Money Economies in Tropical Africa* (New York: United Nations, Department of Economics and Social Affairs, 1955).
UNWIN, G., 'The aims of economic history', in N. B. Harte (ed.), *The Study of Economic History* (London: Frank Cass, 1971) pp. 37–54.
UZAWA, H., 'On a two-sector model of economic growth', *Review of Economic Studies*, 24 (1961) pp. 40–7.
UZAWA, H., 'Optimum technical change in an aggregative model of economic growth', *International Economic Review*, 6 (January 1965) pp. 18–33.
VINOGRADOFF, P., *The Growth of the Manor* (London: George Allen & Unwin, 1911).
WATERMAN, A. M. C., *Economic Fluctuations in Australia, 1948 to 1964* (Canberra: Australian National University Press, 1972).
WHITWELL, G., *The Treasury Line* (Sydney: George Allen & Unwin, 1986).
WICKSTEED, P., 'Co-ordination of the laws of distribution', *Economic Journal*, 4 (1894) pp. 308–13.

WIDDOWSON, E. M., 'Extraction rates – nutritional implications', in Arnold Spicer (ed.), *Bread: Social, Nutritional and Agricultural Aspects of Wheaten Bread* (London: Applied Science Publishers, 1975) pp. 235–46.

WILLIAMSON, J. G., *Late Nineteenth Century American Development: A General Equilibrium History* (London: Cambridge University Press, 1974).

WILLIAMSON, S., 'An interview with Robert W. Fogel', *Cliometric Newsletter*, 5 (July 1990) pp. 26–7.

WRIGLEY, E. A., *Continuity, Chance and Change* (Cambridge University Press, 1990).

WRIGLEY, E. A. and SCHOFIELD, R. S., *The Population History of England, 1541–1871: A Reconstruction* (London: Edward Arnold, 1981).

YOUNG, A. A., 'Increasing returns and economic progress', *Economic Journal*, 38 (1928) pp. 527–42.

Index